Joerg Tremmel is Professor of Intergenerationally Just Policies at Eberhard Karls University of Tübingen, Germany. He was previously Research Fellow at the Grantham Research Institute on Climate Change at the London School of Economics & Political Science. His books include *A Theory of Intergenerational Justice* (2009) and he is Editor-in-Chief of the *Intergenerational Justice Review.*

Katherine Robinson studied public policy at Vanderbilt University, Tennessee and cultural studies at the University of Tübingen in Germany. She is currently with the School of Education, Louisiana State University.

Climate Ethics

ENVIRONMENTAL JUSTICE AND CLIMATE CHANGE

JOERG TREMMEL AND K. ROBINSON

I.B. TAURIS
LONDON · NEW YORK

Published in 2014 by I.B.Tauris & Co Ltd
6 Salem Road, London W2 4BU
175 Fifth Avenue, New York NY 10010
www.ibtauris.com

International Library of Human Geography: 18

ISBN: 978 1 78076 363 7

A full CIP record for this book is available from the British Library
A full CIP record is available from the Library of Congress

Library of Congress Catalog Card Number: available

Typeset by Newgen Publishers, Chennai

CONTENTS

FIGURES AND TABLES

Figures

Table

ABBREVIATIONS

BC	Black carbon
C&C	Contraction & Convergence
CH_4	Methane
CO	Carbon monoxide
CO_2	Carbon dioxide
CO_2e	Equivalent carbon dioxide
EU	European Union
Gt	Gigatonne
GtC^e	Billion metric tons of carbon equivalent
$GtCO_2{}^e$	Billion metric tons of carbon dioxide equivalent
GWP	Global Warming Potential
ICCPR	International Covenant on Civil and Political Rights
ICESCR	International Covenant on Economic, Social and Cultural Rights
IEA	International Energy Agency
IPCC	International Panel on Climate Change
N_2O	Nitrous oxide
NO_x	Nitrogen oxides
OH	Hydroxyl radical
ppb	Parts per billion

ppm	Parts per million
ppmv	Parts per million by volume
UN	United Nations
UNFCCC	United Nations Framework Convention on Climate Change
VOCs	Volatile organic compounds

ACKNOWLEDGMENTS

I would like to thank Ernest Partridge, Thomas Potthast, Peter Lawrence as well as two anonymous reviewers from *Environmental Values* for their valuable suggestions to this book. Stimulating questions were posed both by the students of the Eberhard Karls University of Tübingen at the Forum Scientarium on 7 June 2011, as well as the audience of my presentation at the London School of Economics and Political Science, Grantham Research Institute on Climate Change and the Environment on 18 January 2011. I sincerely thank both groups.

Joerg Tremmel

I would like to thank Joerg for giving me the opportunity to participate in this project, opening my eyes to the world of climate ethics and climate science and for guiding me through the writing and publishing process. I would also like to thank Peter Nagel, Maleen Mack and my family for their persistent questions, encouraging me to constantly re-evaluate the clarity, logic and evidence of the text.

K. Robinson

Morning Dew

Every Morning anew, you refresh the world,
You are always new, never the same.
Epitome of purity and perfection.
A world without you would be a sad world.

PART I

CLIMATE CHANGE: PHYSICAL CAUSES AND EFFECTS

1

INTRODUCTION: WHY AN IMMINENT THREAT STAYS ON THE BACK BURNER

In his book *Our Choice: A Plan to Solve the Climate Crisis*, Al Gore deplores the effect media networks have had on the progress in implementing environmental policies.

> [S]ome opponents of progressive change have grown weary of warnings of catastrophe as a basis of mobilizing support for policy changes. As a consequence, they regularly discount the scientific evidence. This is one of the reasons it has been so difficult to convince civic and business leaders who ought to know better that the climate crisis is not exaggerated.[1]

There is undoubtedly a massive gap between measures recommended based on research results in the natural sciences and those actually implemented; however, it seems implausible that leaders at all levels would fail to act simply because they are tired of hearing about an issue. On the contrary, would this not spur them to act all the more quickly, at the very least so they could stop hearing about it? Given the vast complexity and multifaceted nature of the problem, it seems more reasonable that the 'people who ought to know better' do not simply dismiss scientific evidence as insignificant and irrelevant but instead are put off by its technical nature. Furthermore, it does not typically serve the immediate interests of concerned parties to act on the scientific findings related to climate change. In the minds of most people and countries with the means and influence to affect change, taking measures to prevent or mitigate the effects of climate change implies exorbitant costs, negative economic impacts and disgruntled citizens/customers. None of the catastrophic events predicted by climatologists has come to pass yet, so as the old saying goes – if it ain't broke, don't fix it.

The fact of the matter is, though, that there is indeed something broken – quite a few things, actually. The human species has continually and increasingly adjusted the Earth we inhabit to suit our needs. The rapidity and immensity of our adjustments have upset the delicate balance of the Earth's atmosphere and ecosystems, thus endangering many species, including, to a certain extent, our own. The goal of this book is to present you, the reader, with a consolidated, readable work that makes the issues of climate change accessible, relevant and – well – interesting.

This book is divided into a natural science part and an ethical part and proceeds as follows:

The rest of this introductory chapter discusses the obstacles that prevent the problem of climate change from making it to the forefront of policy agendas, and it lays the groundwork for the next four chapters. There is a brief overview of the development of climate science and general attitudes towards climate change. We introduce the contradiction that exists in the seriousness and urgency of the problem of climate change versus our failure to effectively implement measures that would mitigate it. This sets the stage for the discourse on the science of climate change and its significance in our daily lives as well as in natural disasters and global developments.

In Chapter 2 we outline the science of climate change. Unlike other descriptions of how and why climate change is happening, we avoid technical jargon that makes global warming[2] seem abstract and impossible to understand. The goal of this chapter is to explain climate change in everyday language to which any reader can personally relate. Where technical terms are absolutely necessary, we break them down in a way that makes them as comprehensible as possible to the non-expert. We talk about the substances that cause climate change as well as the physical manifestations of climate change, such as rising temperatures, rising sea level and a higher frequency of some types of natural disasters.

Chapter 3 breaks down the human causes of climate change. We explain first which actions/industries are responsible for what emissions. In the first part of the chapter the substances that cause climate change are tied to their sources, thus providing a concrete context for the 'abstract' substances introduced in Chapter 2. We then select the factors with the highest emissions that are most common in the daily lives of the average person – such as driving/transportation or eating meat – and break down the emissions further. This establishes a material relationship between global warming, the substances that cause them and our part in it.

Climate change is not just about global warming, disappearing species and dying trees. This would be bad enough, but in addition to this, there are very real, very human costs. These costs are already beginning to manifest themselves and are certain to escalate as climate change progresses. Chapter 4 explicates the implications of not addressing issues associated with climate change. These include contagious diseases, sickness, poverty, malnourishment and death. Such consequences are not hypothetical or futuristic. This chapter makes clear that ignoring climate change will result in more problems that will affect us all.

Chapter 5 explores options for and challenges to addressing climate change. The majority of these obstacles have to do with people's mentalities, lifestyles and habits. Ethical aspects are not addressed here in depth; rather the (im)practicality of various adaptations required to mitigate climate change and the costs to individuals. Previous chapters established the reality of climate change and provided ample proof of its existence. Here we first discuss some of the most basic but effective mechanisms that could be employed to mitigate or adapt to the effects of climate change. We then turn to the challenges policymakers face in addressing climate change in the form of the climate sceptics mentioned briefly in Chapter 1. Climate sceptics and fossil fuel industry lobbyists are a significant, if not the main, force hindering efforts to raise awareness of climate change amongst the general public. We first present and refute some of their favourite anti-climate change arguments, then look at their aggressive and sometimes appalling propaganda campaigns. Their refusal to acknowledge the consequences of their actions and the ethical implications serves as a foreword to our discussion of climate ethics in Part II.

Next to the expanding bookshelves in the natural sciences, in international relations and in economics, there is a growing library on the ethics of climate policy. This comes as no surprise when we consider that anthropocentric climate change is one of the greatest problems for the future of mankind. But climate change poses a serious problem for established ethical theories. Gardiner calls the ethics of climate change 'a perfect moral storm',[3] emphasising the especially intricate nature of the problem of climate change for political philosophy, a Pandora's box involving questions of pure distributive justice, international justice, historical justice, intergenerational justice and compensatory justice. There is no dearth of literature on climate ethics breaking down the complexity of the issue, thereby enabling one to arrive at partial conclusions such as: 'Historical Justice demands we do this...' or, 'Intergenerational justice demands we do that...'. Here, in contrast, we attempt to face up to this complexity: that is, to end with a synthesis of the arguments into what we consider to be the most reasonable and the fairest approach to the politics of climate change on a global scale.[4]

In Chapter 6, the object of contention is clarified. The good to be distributed could be a) the capacity of the atmosphere to absorb greenhouse gases, b) the well-being produced by greenhouse gas emissions, c) the economic growth they generate or d) the costs of mitigation and adaptation. This chapter discusses the merits and drawbacks of each perspective. After scrutinising it, it may be said: *greenhouse gas emissions*, or its correlate, the 'absorptive capacity' of the atmosphere should be regarded as the climatically relevant object of distribution.

With an established object of distribution, Chapter 7 asks how atmospheric resources should be divided between the members of present and future generations according to the principles of intergenerational justice. The presently living generations should not fill up the atmosphere with more than 560 billion tons of CO_2e between 2010 and 2050 if a dangerous rise in global temperature is to be avoided. This notion of a safe emissions budget is a restriction for the current generations' consumption of atmospheric resources during their lifetimes. Theoretically, we could arbitrarily set this available budget to 1,000 billion tons and gain more leeway for distribution schemes in the present, e.g., between north and south. This manoeuvre – beneficial for the existing generations, but at the expense of future generations – would contradict what intergenerational justice demands from us. In this chapter we explicate how atmospheric resources can be distributed so as to satisfy the conditions of intergenerational justice.

Keeping in mind our provisions for future generations, how can atmospheric resources be fairly distributed among contemporaries? In Chapter 8, we posit that, according to the maxim of pure distributive justice (without further assumptions), the 'presumption for equality' applies: Everyone gets an equal share of the pie, i.e., every person has the same greenhouse gas emissions budget. For a (static) world population of approximately 7 billion people and a 'safe emissions budget' of 560 gigatons, this results in a per capita emissions budget of 2 tons per year between 2010 and 2050. Dividing the limited budget by the number of individuals corresponds to the ethical stance of emissions/certificate egalitarianism. It grants each person on the planet an equal share of the atmospheric resources. This approach, also called 'Contraction and Convergence', is currently the concept most advocated in ethics. Other distribution schemes, like prioritarianism or the 'difference principle', can't compete with 'Contraction and Convergence'.

In Chapter 9, we explore the concept of international justice, which addresses justice between different countries, regardless of distribution practices within the respective countries. Nations and their political leaders, not individual citizens, make decisions about emission rights in the international arena. Pure distributive justice (see above) could only be applied if we had one world government that had the power to allocate resources fairly among its citizens. In reality, however, the international domain is divided into countries vying for influence and in possession of various bargaining powers, many of whom strictly pursue national interests. The implications of this reality for 'Contraction and Convergence' are discussed in this chapter. A significant part of it is devoted to the question of whether or not population changes are relevant to how emissions rights should be distributed.

In Chapter 10, we discuss what role historical justice, or justice with regard to past generations, should or should not play in the distribution of atmospheric resources. The ability to produce the natural resources coal, oil and gas on a large scale made the Industrial Revolution possible; however, this did not take place everywhere at the same time. Rather, industrialisation arose in various (groups of) countries at different points in history. The cumulative historical emissions of greenhouse gases, i.e., those generated since 1850, are much higher in North America, Europe and the former Soviet bloc than in Africa or Latin America. Based on this, one could support a right to 'overshoot' for low-emission countries. However, several counterarguments can be cited. The weight of these is so considerable that there is no moral *obligation* for the North to compensate the South for initial access to the scarce resource 'atmosphere' (before 1990). The concept of retributive compensation does not apply. Thus, in this chapter we conclude that countries with low historical emissions have no right to 'overshoot', since they were not wronged in any way. They were harmed, but not wronged.

Up to this point we have not addressed the amount of *financial* compensation that the North owes the South. We address this issue in Chapter 11, moving from the currency of greenhouse gases to questions of monetary distribution. We distinguish 'historically founded financial claims' from 'financial claims that are founded with an eye towards the future' and draw conclusions for both sorts.

In Chapter 12, we aggregate the four concepts of justice (intergenerational, pure distributive, international, historical) that have now been applied to the climate problem. This chapter formulates a comprehensive answer to the question: 'What is just with regard to the climate?'

In Chapter 13, we address the question: Can – and should – the language of justice be replaced by the language of rights? 'Rights talk' is used in several ways in the context of climate change. First, it is argued that the present population in developing countries should have a right to greenhouse gas emissions if these emissions are indispensable for a minimum standard of living (so-called 'subsistence emissions'). The hope here is that 'rights talk' can combat poverty in developing countries and reduce inequalities in the living standards between the North and the South.

Second, it is argued that 'rights talk' can help prevent dangerous climate change. Human rights, so the argument goes, provide an effective clarion call for stronger mitigation efforts. Granting such rights to people would impose duties on present people or mankind in general to refrain from emitting too many greenhouse gases. There are two variations of this argument:

a) The position that protecting *present* people's human rights will help stop present emitters from emitting excessive greenhouse gases (in short: 'present people's right to a stable climate' approach).

b) The position that protecting *future* people's human rights will help stop present emitters from emitting (in short: 'future people's right to a stable climate' approach).

We start with the proclaimed 'right to emit of present poor people'. This chapter addresses some key questions of the philosophy of law: What are rights? What is the function of rights? How can rights be justified? What is the relationship between ethical norms, moral rights and legal rights? What do the answers to these questions mean for a proclaimed 'right to emit of present poor people'?

In Chapter 14, we assess the 'present people's right to a stable climate' approach. It seems obvious that climate change has an adverse effect on several already codified human rights, such as right to a standard of living adequate to health and well-being (*1948 Universal Declaration of Human Rights*, article 25) and the right to the highest attainable standard of health (*1976 International Covenant on Economic, Social and Cultural Rights*, articles 12 (1) and (2) (b)), the human right to subsistence (*ICESCR,* article 11) and the human right to life (*1966 International Covenant on Civil and Political Rights*, article 6 (1)). But if this is so, why does climate change goes on unabated? We discuss the juridical difficulties of this human rights approach such as the realisation of the polluter pay principle, the fact that often the complainant's government is not responsible for the harm done and the fact that greenhouse gas emissions are the aggregate result of a myriad of individual activities.

In Chapter 15, we assess the 'future people's right to a stable climate' approach. No one doubts that future generations, once they are present and actual, will have rights, if any of us have rights. But there are some new intricate counterarguments such as indeterminate identities, non-actuality and insecurity. We nevertheless come to the conclusion that no logical or conceptual error is involved in ascribing moral rights to members of future generations.

Chapter 16 discusses if future people can be said to have legal rights. Based on the reasoning in the previous chapter, we answer this question in the affirmative. In fact, some constitutions (e.g. Norway, Japan and Bolivia) *expressis verbis* grant rights to future generations.

Chapter 17 sums up the rights approach and Chapter 18 takes an outlook, pinpointing very recent developments.

Climate change: a problem marginalised by society

Recent events have triggered a new barrage of articles and debates about how climate change should be addressed. The earthquakes in Japan and ensuing nuclear reactor crises in March 2011 prompted Germany to pass legislation to shut down all nuclear power plants by 2022. Consequently, questions of energy sources are at the forefront of politics, as 28 per cent of Germany's electricity is derived from nuclear power (the worldwide average is 14 per cent).[5] While half of the energy supply is to be made up for with wind power, the other half will be replaced by coal power plants and imported nuclear energy. The projected increase in greenhouse gas emissions will almost cancel out the emission reduction targets of the entire EU by 2020.[6] It's easy to see why this decision is controversial. The International Energy Agency's (IEA) projections of emissions increases from 2008 to 2009 made newspaper headlines worldwide, reminding readers of the changes necessary to meet the goals agreed on at the international climate summit in Cancun in 2010.[7] Yet progress continues to be slow – with regard both to people's attitudes and to any strictly implemented practical measures for change. Why?

We present five reasons for the continued failure to adequately address problems stemming from the man-made climate change and expound each in the sections below. First of all, though the climate has been changing as a result of human activities at least since the Industrial Revolution, knowledge and awareness of these changes is new. As such, climate change is for all practical purposes a new problem about which the general public and experts alike are still learning. Second, the field of climate science requires interdisciplinary work across almost all academic disciplines. Biologists, mathematicians, ethicists, sociologists, economists, etc., have to find a common language and have to work together to understand climate change. Third, the highly technical and scientific nature of climate change makes even the simplest, most watered-down explanations of it difficult for a non-expert to comprehend. Fourth, most of the causes and symptoms of climate change are intangible or invisible to the naked eye. How can a person recognise a problem he/she cannot see? Last but not least, the diversity of the affected parties, their priorities and their resources limits the consensus and cooperation that are indispensible in adopting measures to effectively deal with climate change.

Climate change: a new problem

Public knowledge of climate change is relatively new and scientific understanding of it is still evolving. In 1990 the International Panel on Climate Change (IPCC) published its First Assessment Report. This marked a milestone for the field of climate science; one could even argue that the report marked the birth of climate science as its own discipline. By all accounts it was the public declaration from the international scientific research community that they had reached a general consensus that there was sufficient, definitive proof to support the theory of global warming. Since the 1990s, we have been bombarded with information about climate change. It has been the topic of heated debate in the popular media, in academic circles and the political arena. Before this time, knowledge of, or as it were, speculation about, the ways in which humans can affect the climate was limited to a few sporadic studies and articles dealing mostly with hypothetical situations and educated guesses.[8] According to the IPCC report in 2007, 95 per cent of literature on climate change published between 1834 and the time of the report was written after 1951.[9] Scientists are still sorting out the details of how and how fast humans are changing the climate, and the general public is still trying to get its head around the implications of a changing climate. According to a UN Human Development Report, less than two-thirds of the world population is aware of the issues surrounding climate change.[10] Despite the outpouring of information regarding climate change since the 1990s, the understanding of the general population is still lacking due to the complex and interdisciplinary nature of the problem, which brings us to our next challenge in adequately addressing climate change.

Why the problem is so hard to grasp: interdisciplinarity

When we look at the depictions of the science of the process that has been assigned the name 'climate change', the numbers, arrows and symbols belie its true nature: Climate change is an interdisciplinary problem whose almost innumerable challenges refuse to be lumped in or even towards one discipline or area of study. In order to effectively address a problem, one must first understand it. For climate change, this means first understanding the physical properties of the climate, how it is affected by external factors and how it affects the land and people who inhabit it (the gamut of natural sciences). In order to implement effective policies, one must also understand the mindset of the affected population (sociology and political science). All policies must be fiscally feasible (economics) and morally acceptable (ethics and political philosophy). As the second half of this book deals with climate ethics, it is important to emphasise that ethicists are obliged to follow the debates in natural science, engineering, economics, law and international relations before presenting statements on how global warming should be addressed.

Because climate change is so multifaceted, individuals in a position to make changes in favour of saving the environment cannot agree on what changes should be made, who should make them and who should pay the costs of the changes. Meanwhile, the climate crisis barrels on unabated. The technical and vastly complex nature of the problem makes it difficult for the average person to really grasp the subject matter. Government commitments to lower emission levels and climate-saving technologies advertised by big business seem promising. In reality, though, these beacons of hope are often misleading. Many target emission levels are theoretically possible but would require drastic lifestyle changes and sacrifices that most people would not be willing to make. Improvements from climate-friendly technologies are to a large extent unrealistic in the near future because they are either too expensive to be competitive in a market economy or are still in the testing stages of development.

It is not completely fair to vilify industries and governments for misleading the public. Though in some cases the misrepresentation may be intentional, the data that political and business leaders present are usually in and of themselves accurate. Taken out of context, however, these facts are essentially disinformation that results in a pool of knowledge consisting of half-truths. When it comes to climate change, the facts are only of consequence when viewed relative to each other. This, however, requires extensive knowledge in several disciplines. Experts in the aforementioned fields must also consider the ethical implications of the recommendations they make for addressing climate change. Most climate scientists do not have the capacity to absorb and understand the intricacies of all facets of climate change, let alone the average person. There is just too much technical jargon.

The highly scientific and technical nature of the problem

Ironically, the very enormity of the climate crisis prevents most people from being able to comprehend its magnitude and the seriousness of its potential consequences. Take for example one article in a Bloomberg news article: '"Emissions rose the equivalent of 510 million metric tons of carbon...", "Wow", a typical reader might think, stopping mid-sentence at the sight of such a large number, "510 million metric tons sounds like *a lot* of carbon for one year". Then, reading on, "...to 9.14 billion tons in 2010".'[11] Oh. Compared to 9.14 billion, 510 million does not sound like all that much after all. The article goes on to say that global emissions in 2010 were the equivalent of 33.5 billion tons when converted to carbon dioxide. 33.5 billion tons is an even bigger number. And what was converted to carbon dioxide? The carbon emissions? Emissions in general? How can one even convert emissions of one thing into emissions of another? You realise that you have absolutely no concept of what 1 versus 100 versus 1,000 million tons of carbon dioxide is, much less how one would go about converting other greenhouse gases into carbon dioxide.

Translating the terms in which climate change is measured into familiar concepts to which the general population can relate has proved an elusive goal for climate scientists and environmentalists. Fact sheets and flyers likening emissions to everyday terms abound, whether in number of cars driven, gallons/litres of gasoline consumed, electricity consumption, etc. However, a single standard of measurement exists neither for measuring emissions and climate change nor for the translation thereof into more familiar concepts. Conveying climate change and the dangers inherent in it accurately but understandably to the general public is a challenge that scientists face. But it's not only the technicalities of the *problem* that make it difficult to grasp. The complex nature of some of the *solutions* is used by players whose interests and profits lie in environmentally destructive industries, to deceive the public. These parties essentially trick us into thinking that companies are closer to drastically reducing or eliminating greenhouse gas emissions than is the case. They delay the implementation of rigorous measures to mitigate climate change by publicising the advance of research and technology that would enable them to operate normally while eliminating or drastically reducing activities that release large amounts of greenhouse gases. The problem is that many of the emissions-reducing measures referred to are not as far along as those pushing them imply. The difficulty in comprehending such complex processes is further exacerbated by the intangibility of the problem.

Literally not being able to see the problem

The impalpable nature of climate change is a formidable obstacle to adequately addressing the issues. 'Climate' by definition refers to trends and patterns in atmospheric conditions over years, decades or even centuries, whereas 'weather' refers to distinct events in a period of days or weeks. Thus, changes in the climate mean changes in long-term atmospheric norms over time. Few of the symptoms are visible at all. Even those signs of climate change that one can see – more and more dead, brown trees in what used to be uniformly green forests or less ice in the arctic regions – are only easily discernible when looking at before-and-after pictures. Furthermore, symptoms of climate change are often camouflaged as normal deviations from typical weather patterns, adding to its invisibility. As we will see later, it is impossible to attribute individual events to climate change. Distinguishing trends that are indicative of a longer-term change in atmospheric norms from those that are simply normal divergences from typical weather patterns is one of the most difficult and important tasks for climate scientists. Was the 2003 European heat wave a result of steadily rising temperatures all over the world or just a particularly warm summer? Was Hurricane Katrina the beginning of an era of more severe hurricanes and tropical cyclones due to climate change or an isolated extreme weather event? Until definitive patterns emerge that prove a change in the long-term trends we call climate, these remain sporadic extreme weather events. In other words, climate change is happening under our noses, but we can only know for certain in retrospect.

If the manifestations of a changing climate are not readily apparent, it stands to reason that the link between substances destructive to the environment and the actions that produce them will be intangible as well. One could even say that the link between climate change and the actions that cause it is not only intangible but invisible. People cannot see the gasoline from their cars polluting the environment or air conditioners causing global warming, or how destroying trees ultimately results in detrimental levels of carbon dioxide in the environment. These phenomena are perhaps comprehensible theoretically, but without a physical manifestation of changing climate trends, few people are likely to feel an obligation to take responsibility for their actions. Indeed, if there are no visible consequences, there is nothing for which one could feel responsible. Thus we are left with an intangible problem caused by actions whose consequences are invisible. In a fast-paced world where one is bombarded with countless images every day, an invisible problem is easy to ignore – out of sight, out of mind.

As you will see in the coming chapters, though perhaps not conspicuously visible, the consequences of climate change are anything but inconsequential. While virtually simulating the effects of climate change in a way that makes them immediately perceptible to the average person is beyond our capabilities, it may possible to provide a direct analytical link between, say, driving a gas-guzzling car and flooding in Pakistan. It is common knowledge that conventional automobile gasoline, air conditioning and many other everyday activities cause emissions. Also common knowledge is the fact that greenhouse gas emissions are at least partially responsible for current and past natural disasters. Less clear is how these emissions are the same. What does driving less, eating less meat or dressing warmly instead of turning on the heater in Europe have to do with the prevention of hurricanes, heavy flooding or rising temperatures in Asia or North America? The aim of the following chapters is to establish this connection.

Diversity

The last obstacle to addressing global warming that we will present here is the diversity of the concerned parties. Climate change poses new challenges to ethicists, political scientists, economists, scientists of the natural sciences and anyone else who has taken it upon himself or herself to grapple with the issue. The parameters are global, and there is no way to change or influence this fact. We cannot divide the atmosphere into territories around which we build walls for containment and demarcate the borders. Addressing climate change in its entirety necessarily means dealing with experts in all fields, and with individuals and leaders of every society. On the other hand, a one-size-fits-all solution to the issues of climate change is doomed to failure. Measures to mitigate and adapt to climate change have to accommodate local and regional needs and conditions.[12] There is one playing field but many interests and goals. Developing countries are not willing to forgo industrialisation and economic growth, which would be an immediate loss, in order to comply with, say, a global emissions cap in return for a potential long-term gain.[13] From a political standpoint, some experts worry that measures that are too stringent or too sudden would result in rises in the cost of energy so extreme that they would 'trigger strong political protests' in these countries.[14] Even the leaders and experts in developing countries who prioritise long-term environmental solutions over short-term 'quick fixes' cannot ignore the potential negative impacts for citizens if industrial development were to be impeded in favour of environmental policies.

In a globalised world, diversity is an ideal that is celebrated and strived for in communities and societies around the globe. When it comes to climate change, though, it poses a seemingly insurmountable obstacle. In most developed countries, emissions are high and still rising due to the use of fossil fuels and high rates of consumption. These luxuries are highly valued in these countries, though, and citizens are unwilling to moderate their consumption rates in order to reduce greenhouse gas emissions. Furthermore, industrialised countries have more resources to invest in infrastructure to protect them from the dangers of a changing climate. From the perspective of the citizens of these countries and many of their leaders, developing countries should reduce their rates of population growth and industrialise using cleaner technology than, for example, coal plants or traditional cooking stoves that require a lot of energy and are inefficient.

The citizens of developing countries, on the other hand, point to the excessive consumption in industrialised countries and their injudicious use of fossil fuels for energy as the primary causal agent in global warming. Leaders of poor countries say their high birth rates are accompanied by high infant-mortality rates and thus having many children increases the chances that one or more will survive to adulthood and contribute to the household as well as care for elderly parents. Furthermore, inefficient means of cooking and other less than optimal industrial processes are necessary for survival, for which reason it is not fair to demand emissions reductions. Developing countries maintain that their rich, wasteful neighbours should alter their lifestyles, which they could do without risking starvation or poverty.

This is only one extreme example of disagreements between diverse parties. Even within the groupings 'developed' and 'developing' countries, conflicts arise regarding policies that would address climate change. The inability to build a consensus as to how to tackle global warming results in the stalemates often reported by the media as the outcome of international climate summits. Attempting to address climate change as an individual nation is ineffective and the aggregate attempts of many individual nations to address climate change would be too confusing and disorganised. At the end of the day, we are left with a problem of global dimensions and no satisfactory solution.

This precept shakes international diplomatic traditions and methods to the core, not to mention national democratic systems. How we deal with this matter will be the legacy we leave behind and the future of the world ahead. In *Our Choice: A Plan to Solve the Climate Crisis*, Al Gore opines that future generations would be justified in deeming us criminals if we opt for adaptation in the future instead of action in the present.[15] Others are less harsh in their assessment, some even of a completely different opinion. This discord is only further proof of the fact that climate change will be the issue that ultimately unites the world in working towards a common goal or the unsolvable problem that tears apart the tenacious alliances currently in the making. Caught up in the turbulence of political power-plays, beyond comprehension because of its complexity and obscured by its physical intangibility, how could the climate crisis ever make it to the forefront of politics or matters of public importance? It is for this reason that we decided 'yet another' book on climate change is in fact sorely needed. The goal of the following chapters is to explain climate change in terms that are graspable instinctually.

2

THE SCIENCE OF CLIMATE CHANGE

The Intergovernmental Panel on Climate Change (IPCC) is the ultimate authority on the science of climate change. The IPCC Assessment Reports are the most detailed, comprehensive and objective publications concerning climate change that exist, and they are all available with the click of a button as a PDF-file on the IPCC website. Assessment Report 4 (AR4) is 2,583 pages, not including annexes, tables, figures and glossaries. It is divided into three working groups. Each group has its own report roughly between 800 and 900 pages, with a technical summary and a summary for policymakers thoughtfully inserted at the beginning of each report. For those who, as we would expect, balk at the length of the report – not to worry. Beside the full report is a more manageable 52-page synthesis. Obviously, this was our first stop for the latest on climate change. The problem is that because it is such a mammoth-sized report, it takes years to gather the data. As the reports have gotten successively longer, and accordingly more detailed and comprehensive, they have also been published farther and farther apart. The First Assessment Report (FAR) was published in 1990, the second in 1995, the third in 2001 and the fourth in 2007. The fifth is scheduled to be published in different parts between 2013 and 2014. We do not want to discount the IPCC's work. The amount of data the scientists read, sift through and synthesise is a task of behemoth proportions and such a process takes time. In a field as rapidly evolving as climate change, however, time is of the essence.

In our portrayal of climate change, we try to break the process down into the smallest steps possible without making you feel like you are in a middle school science classroom. In a lot, perhaps most, other explanations of climate change, the process is structured according to gases and substances that cause the climate to change unnaturally. The result is a series of paragraphs saying: '**v** has **w** effect on the atmosphere. It is responsible for **x** per cent of greenhouse gas emissions, is present in the atmosphere for **y** number of years and comes from **z**'. The variables do not stand for single words or numbers but rather long, often complicated cause-and-effect explanations. Considering that there are six substances or groups of substances involved in man-made climate change (and even this number can vary according to how the substances are grouped), at the end of such an explanation your head will probably be swimming in numbers, formulae and interactions that bring back (painful?) memories of high school chemistry. In order to give a true layperson's account of climate change – i.e. for people who are vaguely aware that carbon dioxide is bad and we should conserve energy but do not know much more – we have settled on the following structure for this chapter:

The variables listed above (**v** through **z**) are broken down into two sections. First we will talk about the substances involved in the man-made part of climate change and what effect they have on the atmosphere and climate, i.e., **v** and **w** from above. After establishing the 'what' and 'how' of climate change, we define the units that have become the common standards of measurement for quantifying the effect of anthropogenic emissions on the atmosphere. This lays the groundwork for the next part of the chapter in which we talk about how much of the changing climate can be attributed to each agent and how long these effects are predicted to last, or variables **x** and **y** from above. Though we briefly highlight the general categories of emissions sources (our last variable **z** in the previous paragraph), a detailed description of human activities resulting in damage to the environment and the atmosphere thus leading to climate change comprises the next chapter's subject matter. Finally, we bring all of this information together to expound on the concrete manifestations of climate change on Earth (e.g., higher temperatures, more storms, more droughts, etc.). Off-setting effects, i.e., potential solutions to the climate crisis will first be explored after the causes and effects of climate change have been discussed.

Climate change is basically occurring because the atmosphere has a maximum capacity for absorbing, filtering and regulating everything in the air in order to maintain a certain natural balance and this capacity has been/is being maxed out. Much of the excessive strain on the atmosphere is due to people 'adjusting' nature to suit their needs and wishes. For example, increases in carbon dioxide concentrations in the atmosphere between 1990 and 2007 were higher than increases in 1,000 years in the history of the Earth. Nature simply cannot adapt as quickly as people are changing it. After all, the world as we know it evolved over millions of years, changing to develop an atmosphere and ecosystems optimal to life forms currently inhabiting the Earth. Expecting the atmosphere to be able to cope with the drastic changes of a mere 160 years to its millions-of-years-old foundation would be about like submerging people under water and telling them to grow gills. Even the best of swimmers would eventually drown. So what is the water in the lungs of the atmosphere?

The substances of climate change: what? And how?

It's actually pretty simple. There are six man-made or man-emitted substances that contribute to the unnaturally rapid changing of the climate. These are:

carbon dioxide (CO_2);
methane (CH_4);
nitrous oxide (N_2O), better known as laughing gas;
halocarbons;
black carbon (BC) or soot;
and lastly, all constituting one big group, carbon monoxide (CO), volatile organic compounds (VOCs) and nitrogen oxides (NO_x).

Right, now it's a lot clearer, isn't it? It is important, however, to know/keep in mind that some of the energy that reaches Earth from the sun is radiated back into outer space through various atmospheric mechanisms. Just to give you an idea, these include clouds and naturally occurring aerosols and atmospheric gases.[1] We never even feel this energy, and it has no effect on global temperatures.

Ultimately, the Earth has to reflect the same amount of energy back out of the atmosphere as it takes in.[2] How much energy, i.e., heat, remains in the Earth's atmosphere (what we know as temperature), is where our emissions substances come into play. Four of these substances are what we know as 'greenhouse gases' – carbon dioxide, methane, nitrous oxide and the halocarbons.[3]

You have probably heard of the greenhouse effect, maybe even learned about it in grade school or high school.

The Earth, however, is not a giant greenhouse with a glass roof that reduces airflow in order to retain heat, thus we find this analogy somewhat misleading. However, it does provide a convenient catch-all for the four gases that affect the climate in the same way – by retaining heat. The greenhouse gases function like a blanket. Once released into the atmosphere, they capture energy (heat) that would otherwise radiate out into space, i.e., they keep it in the atmosphere. The thicker the blanket, the more heat it can trap. Thus, more greenhouse gases = a thicker blanket = a warmer earth.

Experts prefer the more scientific term 'radiative forcing' – which quite simply refers to the gases forcing the radiating heat waves (in and out of the atmosphere) out of their natural balance.[4] You could also think of this like a traffic jam. Traffic flows steadily until an unnatural obstacle – say an accident or a broken traffic light – causes them to slow down or stop. Because of this obstacle, the number of cars driving past the traffic light per cycle is much lower. In the case of global warming, a 'heat jam' occurs, in which fewer heat waves can leave the atmosphere in a given time period due to the unnatural (i.e., anthropogenic) greenhouse gases blocking the exit. Figure 2.1 demonstrates in the most elementary manner the basic mechanism described above by which greenhouse gases account for warmer temperatures.

The other substances that have been identified as contributing to global climate change are not considered greenhouse gases, because they don't have the same 'blanket' effect. The first of these non-greenhouse gases that we'll discuss is black carbon. Black carbon actually absorbs solar energy, which basically amounts to a cooling effect. That's great, right? Not really. Black carbon alters the natural balance of the atmosphere in several ways. The ultimate effect of black carbon on the atmosphere and the climate is somewhat difficult for a layperson to grasp, because some black

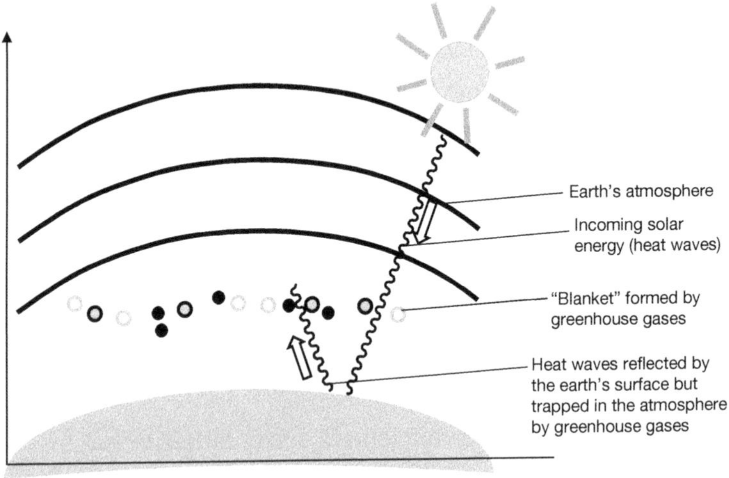

Figure 2.1 How greenhouse gases warm the earth

carbon processes result in warming and others in cooling. Furthermore, the ultimate effect depends on whether black carbon-aerosol compound effects are included in measurements or exclusively direct black carbon effects.

The principle for the warming processes is essentially the same for all mechanisms – black carbon, or soot, absorbs solar radiation, thereby retaining heat that would otherwise have been reflected out of the atmosphere. Initially, black carbon absorbs solar radiation (sunlight) reflected by clouds or the surface of the Earth, thus 'intercepting' solar energy that would otherwise have been on its way out of the atmosphere.[5] Additionally, black carbon absorbs some incoming solar radiation directly.[6] The black carbon that makes it all the way back down to Earth after it has been emitted has to land somewhere, which means raising the temperature of whatever it has landed on (because it has absorbed heat). This is especially important with regard to snow and glaciers, because when black carbon lands on snow or glaciers, it causes the frozen matter to melt (or melt more quickly).

This is not only a direct illustration of rising temperatures due to black carbon, but as you'll see in the coming pages also contributes to a self-perpetuating mechanism that further exacerbates the problem of climate change.[7] Lastly, some soot ends up inside cloud drops and ice crystals causing it to absorb more sunlight, i.e., reducing the amount of solar energy radiated out of the Earth's atmosphere.[8] Black carbon is the only major contributor to global warming that is visible to the naked eye. Its name is deceiving because the black carbon emissions that we can see are actually small brown particles in the air. In cities this takes the form of smog. On snow, it looks like mud and dirt has been mixed in with what should be pristine white. Figure 2.2 below depicts BC's contributions to global warming.

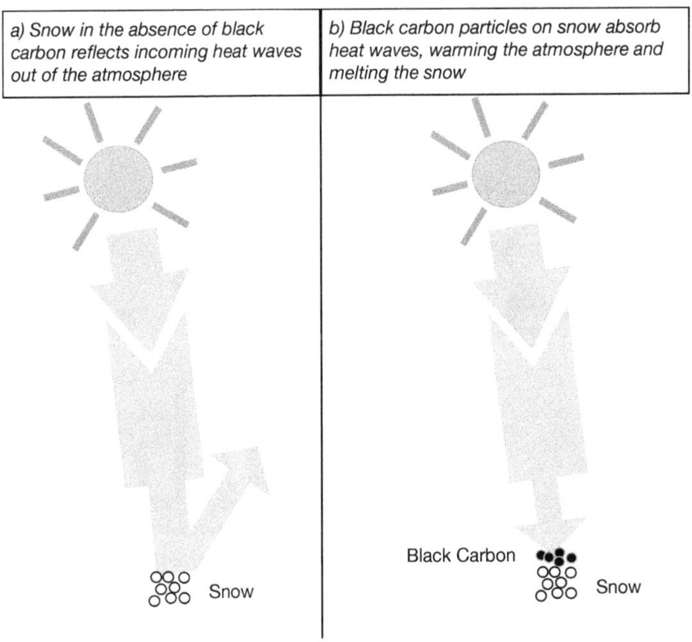

Figure 2.2 How black carbon contributes to rising temperatures

Carbon monoxide and the substances grouped into the category 'volatile organic compounds', or VOCs, are the parenthetical elements of factors responsible for climate change, because their presence in the atmosphere alone has no impact on the climate or atmospheric balance. Because of this, the effects CO and VOCs have on the environment are classified as 'indirect' effects. Accordingly, the means by which these compounds affect the climate are considerably complex, even when broken down to the most basic of terms and processes. For example the picture below is taken from an interdisciplinary book on climate change. The section of the chapter from which this excerpt comes is 'aimed especially at the non-expert'.[9] Figure 2.3 below, the authors say, 'reviews, with much simplification, the chemical reactions involved'.[10] If this graphic is supposed to be simplified for the non-expert, it may have missed its mark.

For our purposes, it suffices to say carbon monoxide, nitrous oxides and VOCs break down quickly in the environment

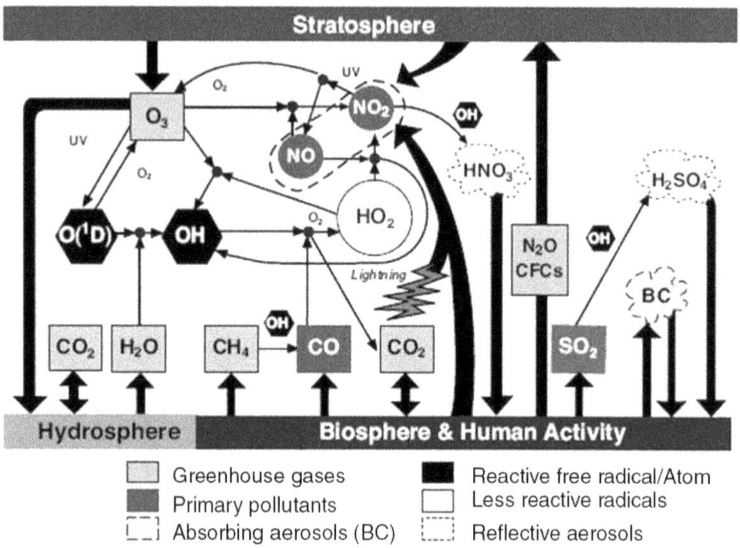

Figure 2.3 'Simplified' illustration of how CO and VOCs affect climate change

and 'recombine' into substances that are harmful to the environment.[11] Sometimes it's in the form of carbon dioxide or methane, which means more of these greenhouse gases in the atmosphere than from direct emissions alone. Other times, the resulting compounds are climatically innocuous; however, the CO and VOCs also react with a compound that plays a huge role in breaking down/absorbing methane in the atmosphere, thus effectively eliminating hydroxyl radicals (OH) which are a major sink for one of the most potent greenhouse gases.

Which substances are responsible for how much of climate change?

Now that we have a basic idea of which substances damage the environment and how, the next question is, 'How much climate change can be attributed to each element?' This

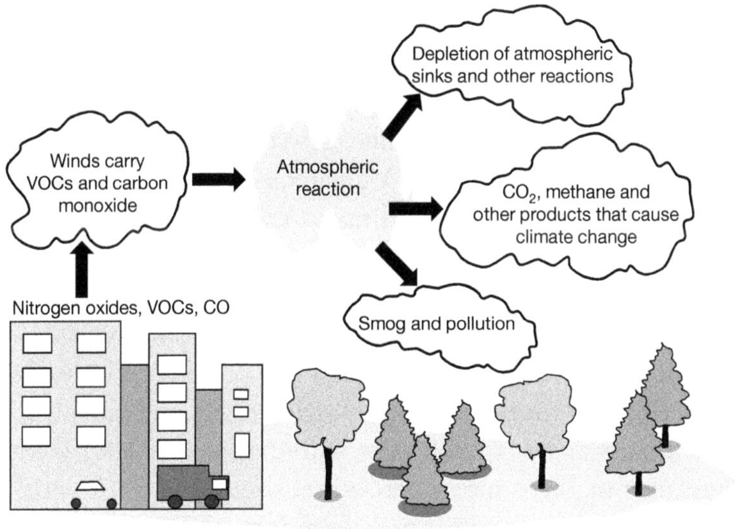

Figure 2.4 How CO and VOCs intensify climate change

is important in deciding where to focus efforts to reduce emissions, i.e., identifying the most effective and efficient measures to mitigate climate change. Understanding which substances are responsible for how much climate change and/or trying to decipher tables or graphs illustrating this relationship is possibly the most complicated aspect of climate science for a layperson. Even the most rudimentary literature on climate change tosses technical terms around, which creates a further disconnect between individual actions and global climate change. Thus, before going into more detail, it is necessary to translate some of the concepts used in discussions on the environment into everyday language.

First of all, there are several standards by which emissions are measured. They all ultimately amount to the same changes in our environment, but the array of terms and numbers can be daunting to someone new to or unfamiliar with the field. Depending on the source, the factors and consequences of climate change may be discussed in terms of gigatons, parts per million (ppm)/parts per billion (ppb), ppm or ppb in carbon dioxide or carbon dioxide equivalents, radiative forcing, global warming potential (GWP), global temperature changes, watts per square metre, ...and the list goes on. For example, one newspaper article explains that we currently release 50 gigatons of greenhouse gases into the atmosphere. According to experts in the field, it continues, we must reduce this amount to 44 gigatons by 2020.[12] Other news articles talk about parts per million of carbon dioxide in the atmosphere. Furthermore, sometimes emissions are discussed in terms of carbon dioxide only and other times in terms of carbon dioxide emission equivalents for all substances. In the next section, we will clarify the meanings of these terms and establish a basis from which to explicate the concrete manifestations and consequences of climate change. Since climate change texts aimed at the general public talk most often in terms of ppm, ppb, gigatons and carbon dioxide equivalents, we will limit our discussion to these terms. Explicating all the measurements used in climate science is beyond the scope of this book.

Measuring climate change

When discussing climate change, technical literature and popular media articles alike often refer to how many parts per million (ppm) carbon dioxide are in the atmosphere. This is a measure of the ratio of carbon dioxide particles to the total number of particles of dry air.[13] In 2011, for example, for every million particles in the atmosphere, 391.57 of them were carbon dioxide, i.e., 391.57 ppm CO_2.[14] One in one million sounds like a laughably insignificant number. Imagine, however, that these particles were cancer cells. In the early stages of cancer, the number of cancer cells compared to the number of healthy cells in a person's body is relatively low (about 1 in 10,000). However, because these cells are destructive to the body, when they begin to multiply more rapidly than the immune system can destroy them, they wreak havoc on the body and can result in death. In the same way, though moderate amounts of CO_2 are necessary to our survival on Earth, excessive emissions of carbon dioxide particles and the other particles causing climate change are destructive to the atmosphere in ways that disrupt the lives not only of humans but of almost all species inhabiting this planet.

Methane and nitrous oxide are measured in parts per billion, an even smaller ratio, because there is less of these gases in the atmosphere than there is of carbon dioxide. That said, one particle of methane or nitrous dioxide is 'worse' for the environment than one particle of CO_2. This will become clear after our discussion on carbon dioxide equivalents and the changes in greenhouse gas concentrations in the atmosphere caused by people.

The second term that one is likely to come across when reading news articles or the like about climate change is gigatons. This word may be written as gigatons or billions of tons. One gigatons is approximately equal to twice the weight of all the people living on Earth. Most people cannot even begin to conceive such a large mass, much less one that is invisible to the naked eye yet present in the very atmosphere in which we live. While such weight is difficult to equate to more familiar terms, it helps to remember the size of the Earth's atmosphere. The 'upper edge' of the atmosphere, so to speak, or its approximate border with outer space, is 99.9 kilometres (62 miles) above the surface of the Earth.[15] So much 'storage room' makes it possible for us to send masses of greenhouse gases many times the weight of all of humanity into the atmosphere each year without perceiving the additional tonnage of our emissions.

Now that we have an idea of the terms in which climate change substances are measured, it is time to look at how much climate change each substance causes.

A common currency: carbon dioxide equivalents

Laying out which substances are responsible for 'how much' climate change is not as simple as just displaying the amount of emissions per substance in a pie graph. First of all, each gas or substance affects the atmosphere to different degrees. One ton of methane, for example, would have the same effect as 25 tons of carbon dioxide in the atmosphere in 100 years, but the effect of 72 tons of CO_2 in 20 years.[16] Consequently, raw numbers or percentages of climate-changing substances are largely meaningless. For a more straightforward measure of comparison, those in climate-related fields now often refer to carbon dioxide equivalents. Carbon dioxide equivalents are exactly what they sound like: how much carbon dioxide a given amount of a different greenhouse gas equals. You can think of this as the classic ton of bricks/ton of feathers comparison. Obviously, both a ton of bricks and a ton of feathers weigh one ton. The ton of feathers, though, would have many more individual feathers than the number of bricks in a ton of bricks. Finding CO_2 equivalent emission is like determining how many feathers you would have to throw together to travel the same distance as one thrown brick, only you are finding out how much carbon dioxide would have to be emitted to have the same effect of emitting, say, 1 ton of methane or nitrous oxide over 20 or 100 years.

The IPCC uses two basic carbon dioxide equivalents: the carbon dioxide equivalent concentration and the carbon dioxide equivalent emission. A carbon dioxide equivalent *concentration*, or equivalent carbon dioxide (CO_2e), is the *concentration* of carbon dioxide that would have the same effect as the concentration of another greenhouse gas or mixture thereof.[17] Taking our feather and brick example, this would be like finding out how many feathers it would take to exert the same amount of pressure as one brick on a given surface area. The equivalent carbon dioxide of a gas is measured in parts per million by volume (ppmv) or as a decimal in terms of radiative forcing (RF).

A carbon dioxide equivalent *emission* is the amount of CO_2 that would cause the same warming effect, or radiative forcing, as the emission of a different greenhouse gas over a given period of time.[18] These are measured in gigatons of carbon (GtCe) or gigatons of carbon dioxide ($GtCO_2e$) because it refers to the amount of weight of the gas that is emitted. This is also called a gases global warming potential (GWP). For example, for a 20-year period, 1 gigaton of methane would have the same effect as 72 gigatons of carbon dioxide.[19] Therefore, its GWP or $GtCO_2e$ in 20 years is 72. When talking about global warming potentials it is important to be aware of the given time span. Since greenhouse gases break down in the atmosphere over time, the longer the time span is, the lower the GWP will be (except for carbon dioxide, of course, since it is the reference point). On a 100-year time scale, methane's GWP is 25, on a 500-year time scale it is 7.6 and so on.

In the next section we will briefly discuss the amount of climate change substances in the atmosphere, their contribution to climate change and give an overview of the sources from which they are emitted. Figure 2.5 shows the changes in concentration of three major greenhouse gases from the year 0 to 2005. This is helpful in putting into perspective the effect humans have had on the makeup of the atmosphere. The data are taken from ice-core samples, that allow scientists to determine atmospheric content in the past as

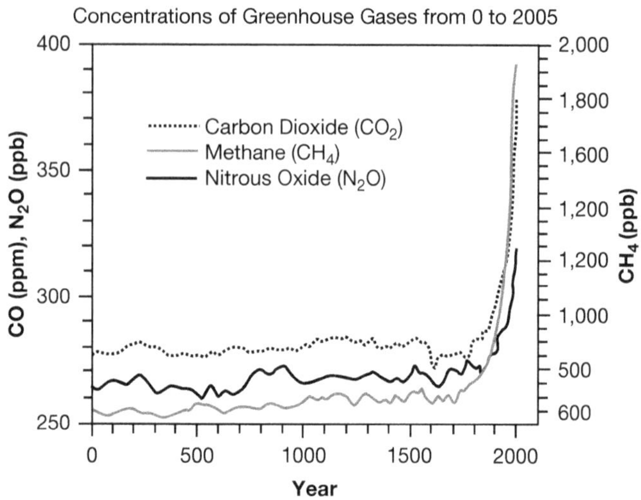

Figure 2.5 Changes in CO_2, N_2O and CH_4 concentrations from 0 to 2005

well as for more recent years. As you can see, until the late eighteenth/early nineteenth century, atmospheric concentrations of carbon dioxide, methane and nitrous oxide were relatively stable. With the onset of the Industrial Revolution, however, humans began to drastically change their environment without even knowing it.

Sources of anthropogenic emissions and their proportional contributions to climate change: an overview

Carbon dioxide has caused the most warming of all the greenhouse gases since 1750.[20] In 2004, it constituted 77 per cent of all anthropogenic greenhouse gas emissions.[21] According to the IPCC, the bulk of anthropogenic carbon dioxide emissions occur as a result of either the use of fossil fuels (e.g., in transportation, heating and cooling and the industrial production of cement and other products) or deforestation (e.g., land-clearing for residential living space or cutting down trees for firewood, paper production or other material).[22] Carbon dioxide concentrations in the atmosphere have increased from 280 ppm in pre-industrial times to an average of 391.57 ppm in 2011.[23] Figure 2.6 shows CO_2 emissions since 1850 in tonnage. While steadily rising since the Industrial Revolution, CO_2 emissions shot up in the second half of the twentieth century. This trend is consistent across all anthropogenic climate change substances as concentrations that had been accumulating since the nineteenth century and ever-rising emissions rates began to overwhelm the atmosphere.

Methane is the second largest cause of global warming after carbon dioxide.[24] It remains in the atmosphere for 8.4 years after is emitted, a relatively short lifetime compared to carbon dioxide, but as we mentioned above, it has a global

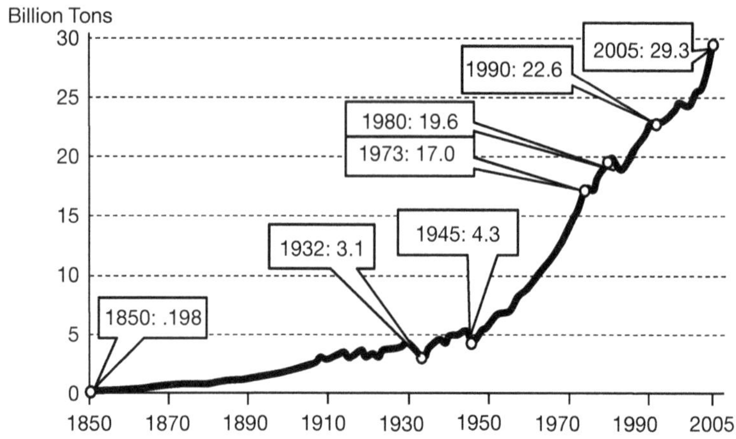

Figure 2.6 Global CO_2 emissions since 1850

warming potential of 72 over 20 years and is therefore much worse for the environment than carbon dioxide. Human-caused methane emissions result mostly from rice agriculture, livestock, landfills and waste treatment, some biomass burning, and fossil fuel combustion.[25] As you can see in Figures 2.5 and 2.7, methane emissions have increased dramatically in the last few centuries. In pre-industrial times, the atmospheric concentration of methane was 715 ppb; by 2005, it had risen to 1,774 ppb, or almost 150 per cent more concentrated than it had been.[26] Compared to the 35 per cent increase in carbon dioxide concentrations in the same period, this number seems shocking. One mitigating factor, though, is that because methane's emission growth rates have fallen since the 1990s, concentrations in the atmosphere are not currently increasing.[27] As can be seen in Figure 2.7, methane rose dramatically in the second half of the twentieth century, though one can see that the tip of the curve began to tail in the last couple of decades of the century.

Nitrous oxide is the third and last of the 'major' greenhouse gases.[28] It is the only major greenhouse gas for which

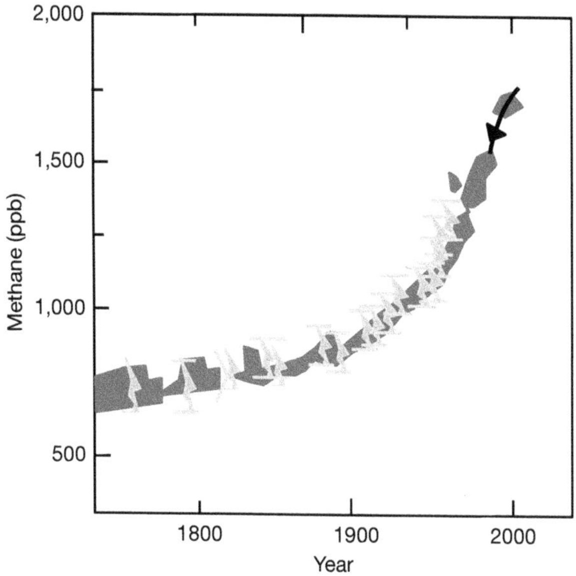

Figure 2.7 Change in ppb methane in the atmosphere since 1800

anthropogenic emissions are approximately equal to natural emissions, though concentrations of nitrous oxide in the atmosphere grew almost two times higher than average between 1960 and 1990.[29] This trend is evident in Figure 2.8 below and looks similar to the curves of the other greenhouse gases over the same time period. Though nitrous oxide is the greenhouse gas with the lowest concentration in the atmosphere, it is also the greenhouse gas with the most intense effect on climate. Once emitted, it remains in the atmosphere for 114 years and has a GWP of 298 on a 100-year timescale.[30] Concentrations of nitrous oxide 'only' rose from 270 ppb to 319 ppb from pre-industrial times to 2005,[31] but its high global warming potential makes this increase significant. Nitrous oxide is emitted from the fertilizer of agricultural soils, from burning biomass, from raising livestock and from some industrial processes like the production of nylon.[32]

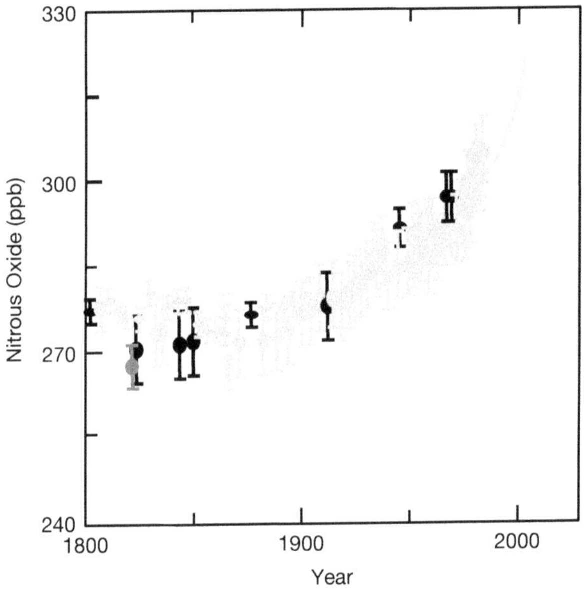

Figure 2.8 ppb nitrous oxide emitted since 1800

In the last section we talked about how the last two (groups of) substances that cause climate change – black carbon and carbon monoxide and volatile organic compounds – are fundamentally different from greenhouse gases in how they affect the atmosphere. They also differ in the extent of their contribution to climate change as well as how long their effects last. The exact contribution of black carbon to global warming is a contested subject. Black carbon affects the environment in many ways. As we have seen, carbon absorbs heat waves when it lands on otherwise reflective surfaces, but floating around in the atmosphere it can also act as a cooling element. Regardless, there is no doubt that black carbon pollutes the air and disrupts snow and ice cover trends. Its atmospheric lifetime is only a couple of weeks and improvements in air quality are noticeable even to the general population. Black carbon is a big problem for poor countries whose citizens use old-fashioned stoves for cooking that send up thick clouds of smoke, filling the air with black carbon.

The last agents of climate change we deal with here are carbon monoxide and volatile organic compounds. CO and VOCs only remain in the atmosphere for between a few days and a few weeks.[33] The atmospheric lifetime of climate-changing substances has huge political implications. The longer a gas is in the atmosphere, the farther it can travel; i.e., the more regions it can reach. CO and VOCs mix and break down quickly, so these compounds don't travel very far. In other words, there is no 'global distribution' of the pollution. CO and VOCs remain confined to the regions/hemispheres (Northern or Southern) in which they are emitted.[34] Since their part in climate change is different from that of greenhouse gases and because they are in the atmosphere for such a short period of time, we do not talk about carbon monoxide and VOCs in terms of GWP. Their threat to the climate is immediate and as such CO and VOCs are one of the few climate change factors for which, when emissions levels are reduced, environmental improvements are measurable in the short term.

In the next section, we discuss how much of all of these substances the atmosphere can tolerate, or more accurately, how much atmospheric change our planets current inhabitants can tolerate before their well-being and/or livelihoods are threatened by climate change.

Atmospheric limits

Atmospheric sinks were briefly mentioned above with regard to the substances that cause climate change. Before we get into detail about targets, reductions and limits, it is important to have a general idea of how the atmosphere interacts with greenhouse gases and other climate-changing substances. These processes are the central factor in the determination of necessary limits and reductions of emissions harmful to the atmosphere.

There are several ways that nature breaks down and absorbs the substances that we emit into the air. Each of these mechanisms is referred to as a sink, because the atmosphere acts as a large basin. We are filling the basin with the by-products or waste from our everyday and industrial processes. How much of any given gas or substance a sink can take before it 'spills over' is then the limit for that substance for that sink. If we fill up all of the sinks and continue to emit gases into the atmosphere, the climate will change drastically and Earth will experience irreversible changes. This concept is best illustrated with the help of an analogy:

> Imagine a large sink in which all of your food scraps are thrown. The sink is surrounded by animals that eat from the bowl. Each animal eats different amounts and types of food. The bowl is never empty, but the animals ensure that it never overflows. The amount of food thrown out is balanced by the amount eaten by the animals. If you have a party, the scraps are thrown into the same sink. The animals continue to eat, but the amount of food in the sink is more than before because of the party. If you continue to have parties, the animals will not be able to keep up. The sink slowly begins to fill. You are not limited in the amount of food you make and throw away, but the animals are limited by their body size and diets as to how much they eat. The sink, of course, is limited by its physical capacity to hold food. If it is not eaten, some of the food will decay naturally, but this takes a long time. Eventually, the sink spills over and food scraps begin to fill the space around it. If you cannot clean it up,

> you have to wait and hope the animals will catch up and not be drowned in the food. Until then your quality of life and living space are reduced by the growing amount of trash surrounding you.

In our analogy, each animal's stomach capacity represents a sink for anthropogenic emissions. The 'party-phase', which upsets the balance between discarded and consumed food scraps, represents industrialisation and our reliance on fossil fuels thereafter. The largest sinks for emissions harmful to the atmosphere include the ocean, forests, non-forest plant life and atmospheric substances that react with or break down greenhouse gases to render them harmless to the environment. There is no absolute limit *per se* for substances that cause climate change in the atmosphere. Let's go back to our party-sink-animal analogy. Obviously, not all people are the same. Thus, the individuals at the party would have different tolerances for the amount of trash that could spill over the sink before it becomes destructive to them. This tolerance would be based on individual needs, preferences, medical conditions, age and proximity to the sink. In the same way, the species and groups of people on the Earth have different tolerances for degrees of climate change. Thus, the point at which the anthropogenic climate change is deemed dangerous and/or unacceptable depends on where and how a person lives as well as his/her priorities and values.

That said, climate change is still a global problem and in order to address it effectively, international agreements have to be made that establish limits and targets concerning emissions and climate change. To this end, so-called 'tipping points' have been identified that absolutely must be avoided. This term has been a topic of much discussion and has many definitions in scientific literature on climate change. In the context of the anthropogenic climate change and this book, tipping points refer to when emissions from human activities have resulted in irreversible and self-accelerating changes in natural elements or phenomena. The tipping elements in question include major ocean currents that play a significant part in regulating the climate, degradation of rain forests, the melting of the Greenland ice sheet, permafrost and arctic sea ice and shifts in monsoon trends.[35] The limits we discuss in the next paragraph would theoretically avoid these critical points that may make it impossible to mitigate and/or adapt to climate change, i.e., avoid dangerous climate change.

In order to prevent some of the most dangerous consequences of climate change, we have to limit the global temperature rise to 2°C (3.6°F) relative to pre-industrial levels.[36] According to the United Nations Framework Convention on Climate Change (UNFCCC), '[a] 2 degrees Celsius/Centigrade is the highest rise we can afford if we want a 50 per cent chance of avoiding the worst effects of climate change.'[37] Limiting the rise in average temperatures to 2 degrees, however, is becoming an increasingly far-fetched target. In terms of greenhouse gas concentrations, this means that we could not exceed 400 ppm CO_2 or 450 CO_2e[38]; however, we are already dangerously close to these concentrations and emissions growth rates continue to rise. Targets seem farther and farther out-of-reach, and some experts are suggesting that the only way to achieve these goals at all is to initially overshoot 'climate-friendly' levels and then cut emissions so as to create a negative-emissions world.[39] Emissions levels would have to peak by 2015 then fall drastically so that atmospheric sinks could break down more substances than we emit, thus slowly restoring the atmosphere's natural balance. This, however, does not guarantee that the tipping points we discussed above would not be reached. Theoretically, upper limits would only be exceeded for a short time and then drop back to 'safe' levels before tipping points have been reached. But given the uncertainty concerning climate change's exact effect on natural systems and elements, we cannot be sure that the tipping points will not be reached in the interval in which atmospheric concentrations are highest. After all, as we will see in the next section, there is already irrefutable proof that climate is changing, and it cannot be a coincidence that these changes coincide with and are proportional to anthropogenic emissions.

Manifestations of climate change

'Warming of the climate system is unequivocal,' states the IPCC in its Assessment Report Four.[40] Indeed, just as dramatic upward trends in atmospheric concentrations of greenhouse gases depicted in Figures 2.5–2.8 above are unmistakable, so too are changes in various climate elements. In this section, we talk about how climate averages are already change and, based on this, how these trends are likely to unfold in the future. Figure 2.9 below shows changes in average temperature, sea level and snow cover between 1850 and 2000 or from the time at which data became available. The left side of the vertical axis shows the amount by which averages have risen or, in the case of snow cover, fallen while the right side of the vertical axis shows the averages themselves. Though average temperature and sea level have risen since 1850, the values on the left side of the y-axis are negative up to the mid-twentieth century because averages are relative to mean temperature and sea level from 1960–1990. For example, the change in average temperature for 1850 is graphed at –0.3°C, thus 1960–1990 averages were 0.3°C higher than 1850 temperature averages. All in all, average temperatures on Earth rose from

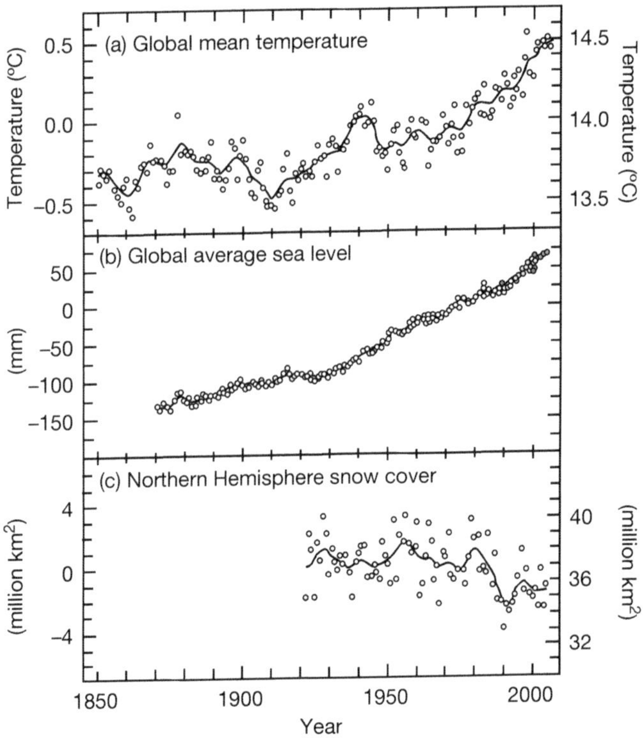

Figure 2.9 Changes in temperature, sea level and snow cover (1850–2000)

13.7°C (56.66°F) in 1850 to 14.5°C (58.1°F) in 2000. In the same time period, sea level has risen by 0.2 metres or almost 8 inches. Data for average snow cover in the Northern Hemisphere is only available starting in 1920.

Rising temperatures

Rising temperatures are by far the most talked about consequence of climate change. Global average temperatures have risen by 0.75°C (1.35°F) since 1900.[41] With a goal of limiting the rise in average global temperature to two degrees, we have 1.25 degrees, or over half of our budget, 'left to spend' so to speak. Based on past trends and possible future scenarios, preventing the average global temperature from rising by another 1.25°C is a lofty goal. As you can see in Figure 2.10 taken from the 2011 Human Development Report, the trend in the average world temperature from 1900 to 2008 changed dramatically in the last few decades of the twentieth century.

This graphic is particularly useful because it shows not only the average rise in temperature in the course of a century but also separately graphs the variations in average annual temperature. Plotting both average temperatures on the same plane explains why global warming does not always feel so warm. There are and will continue to be years with extremely cold winters (relative to average winter temperatures for any given region). These cold years, however, will be fewer and farther between, and they might not be as cold as cold years 50 years before. People often have short-term memory when it comes to weather. Thus, one or two or even three very cold winters will feel like 'proof' that the

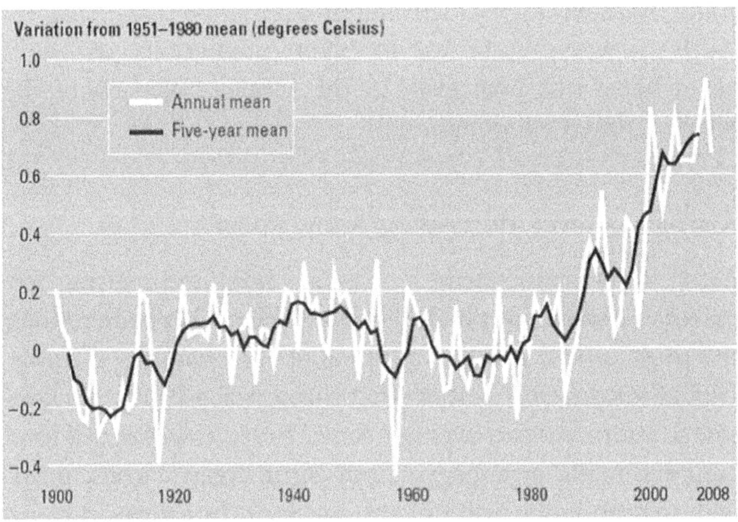

Figure 2.10 Changes in the average world temperature from 1900–2008

earth is not getting any warmer. Figure 2.10 below shows very clearly, though, that it is indisputable that temperatures are rising despite fluctuations in annual averages.

In order to get ahead of potential problems associated with higher global temperatures, scientists around the world have been working to develop models that can essentially predict the future. Criteria factored into these models run the gamut from economic to social to environmental and even political indicators for outlooks for the future. Since 1990 the IPCC has been compiling data based on past and current trends as well as theoretical models in the scientific literature to develop various future scenarios and explore their implications for mankind.[42] The scenarios are basically a long series of 'if... then' statements. Based on social, economic and technological developments, scientists predict emission levels and their impact on the environment. Figure 2.11 depicts potential changes in global average temperature up to year 2100. Values are relative to temperature averages between 1980 and 1990. If you remember from above, we have 1.25°C 'left' in our budget for rising temperatures; however, according to expert predictions, we could not achieve this goal even in the idealistic version of the future painted by scenario B1.

Rising sea level, decreasing snow cover

After rising temperatures, rising sea level and melting ice/decreasing snow cover in the Northern Hemisphere are among the most talked about symptoms of the changing climate. Global measurements show that through the 1900s, sea level rose 1.7 mm/year on average. Since 1993, however, sea level has been rising by approximately 3 mm/year.[44] These numbers may sound small and insignificant, but keep in mind we are talking about 3 mm more water in all of the oceans

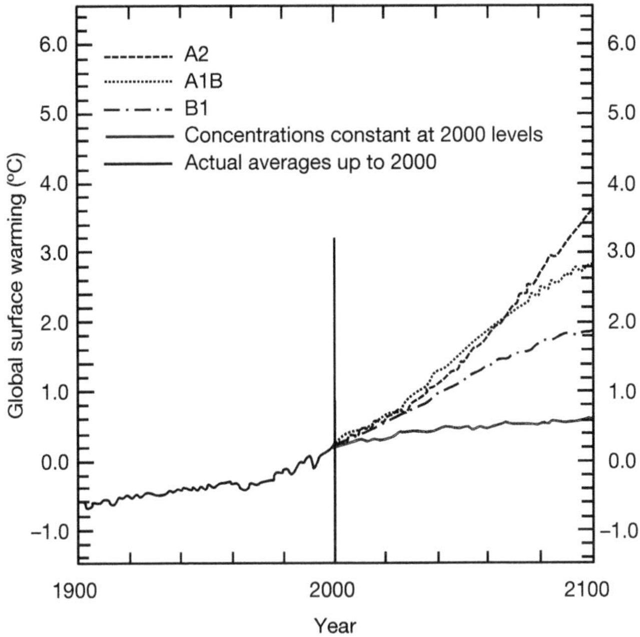

Summary of scenarios:

A2: heterogeneous by region, growing global population, economic growth and technological change slow compared to other scenarios.

A1B: rapid economic growth, global population declining after mid-century, rapid development of more efficient technologies, sources of energy are balanced between fossil fuel intensive and non-fossil fuel intensive.

B1: global population declines after mid-century, service and information-oriented economy, clean and efficient technologies, global solutions to social, environmental and economic problems.

Figure 2.11 Potential temperature changes through 2100[43]

on Earth. That is a lot of water. Furthermore, though small in absolute terms, this 76 per cent increase in the rate of change is over a very short period of time, and this number is expected to continue to rise.[45] For the scenarios used in Figure 2.11 for potential changes in average temperature through 2100, the corresponding rises in average sea level predicted by the IPCC are between 0.23 and 0.51 metres for scenario A2, 0.28–0.48 metres for A1B and 0.18–0.38

for B1. This is a further observable, measurable, irrefutable proof of climate change. It is also easy for the average person to understand. And very dangerous. Should the sea level rise by 0.5 metre, an area the size of France and Italy combined (almost 1 million square kilometres) would be flooded and 170 million people would be affected.[46] As we will see in Chapter 4, this would represent an enormous cost in lives and well-being. For this reason it is important to understand how and why sea level rises in order to implement measures to mitigate or adapt to higher water levels.

There are two major causes of a rise in the sea level:

(1) ocean water expanding as it warms (thermal expansion);
(2) water that results from the melting of what was previously land-based ice.[47]

It is not just its usefulness as a 'user-friendly' illustration of a warming world that makes rising sea level so central in discussions of climate change: increasing water levels are not only a result of but also propel global warming. This is one of the tipping points we discussed when talking about emission limits necessary to preserve the balance of the atmosphere and the climate. Put simply, rising sea level is the beginning of a vicious cycle – or, in this case two vicious cycles. When glacier ice melts, it becomes water. Ice and snow are reflective surfaces and thus reflect sunlight out of the atmosphere which results in cooler temperatures. Water, on the other hand, is an opaque surface that absorbs sunlight, affording warmer temperatures. Warmer temperatures cause ice to melt, eliminating another reflective surface and adding an absorptive surface for sunlight, and our cycle is in motion.[48] Additionally, northern regions are seeing less snow as warmer temperatures cause more precipitation to fall as rain than in the past.

Changes in precipitation patterns

Weather extremes and precipitation patterns are affected by global warming; scientists have proven and documented a shift in the amount, type and frequency of various weather events since 1990. For example, as a result of Earth's rising temperatures, both droughts and flooding are becoming more frequent and more severe. This might seem counter-intuitive since these two weather phenomena are polar opposites of each other. This paradox, however, once explained, is actually quite easy to imagine. In the preceding paragraph, we laid out how and why sea level is rising. This phenomenon also contributes to changes in precipitation. Higher atmospheric temperatures and melting ice obviously mean higher water temperatures. These persistently higher water temperatures as a result of climate change are contributing to changes in precipitation and other weather patterns. How? As water temperatures rise, so do water vapour amounts – ice melts to water, water evaporates into water vapour.[49] Because most precipitation is a result of weather systems feeding on the water vapour in the atmosphere, more water vapour means more precipitation.[50]

Drought occurs under exactly the opposite conditions with the exception of rising temperatures. As temperatures get warmer, land without significant moisture dries more quickly, i.e., there is less water vapour in the atmosphere. Less water vapour results in less precipitation thus increasing the risks and severity of drought.[51] What makes this trend even more problematic is that it exacerbates already existing challenges. Rising global temperatures are essentially causing dry areas to become drier and humid areas to become more humid, thus increasing the risk of drought in regions already arid and suffering from lack of water *and* increasing the risk of floods in areas that struggle regularly with inundation. For example, Sub-Saharan Africa, which already has a dry, desert landscape, has experienced the sharpest drop in average rainfall over the past few decades. In this region, average rainfall decreased by 7 mm per year between 1951 and 1980.[52]

The occurrence of weather extremes – tornadoes, hurricanes, monsoons, etc. – is dictated by complicated interactions of wind and water currents and other atmospheric factors. Because of this multitude of variables, it is impossible to say if the changing climate is responsible for any given extreme weather event. Even in a stable climate, extreme events are, to some extent, 'normal'.[53] Scientists can predict, though, how the frequency and intensity of certain weather phenomena could be affected by changes in the climate caused by people. Tropical cyclones, for example, would have higher wind speeds and heavier precipitation. As proof of this phenomenon, scientists point out that the number category 4 and 5 hurricanes (5 being the highest number on the scale of hurricane categories) increased significantly between 1977 and 2007.[54] In a warmer climate, extreme weather events will be more concentrated. The effect is analogous to the intensity of water flow from a shower head. At a normal setting, the water flows with equal pressure from all openings. Adjusting the shower head to increase water pressure causes water to flow more intensely out of a few openings while the others remain dry. Global warming essentially turns up the pressure in the Earth's shower – precipitation is concentrated into more intense events with longer dry periods in between.[55]

Some extreme weather events like tsunamis happen completely independent of climate change. That is, higher temperatures, changing precipitation patterns and rising sea level do not affect the frequency or severity of their occurrence. This does not mean, however, that the damages caused by these catastrophes are independent of climate factors as well.

For example, the changing climate has no bearing on the frequency or severity of tsunamis. It is, however, affecting coral life in the regions where tsunamis hit. Corals are sensitive to their environment, and because of rising sea levels and water pollution due to climate change, many corals are dying. How is this related to tsunamis? Coral reefs are an essential line of defence against tsunamis – they absorb some of the impact from the floods and thus mitigate the damage on land. Thus, fewer corals mean more severe flooding and accordingly destruction in affected regions. We will now look at what factors play the biggest roles in the anthropogenic climate change.

3

THE CULPRITS OF CLIMATE CHANGE

Scientists and experts in the field of climate science use the phrase 'culprits of climate change' to refer to the gases that cause it. While catchy, the name in and of itself does little to create a concrete image of the phenomena to which it refers. For this reason, we prefer to apply the label 'culprits of climate change' to the activities that result in the release of the substances we talked about in Chapter 2 into the atmosphere. People's consumption patterns together with rapid population growth are the driving forces of anthropogenic climate change. Consumption patterns are the ways in which people fulfil their needs. This includes the means of transportation by which one gets from point A to point B, the amount of living space one occupies, what and how much a person eats and what and how much a person buys (clothes, furniture, luxuries, etc.), just to name a few.

First, we will chronicle the activities that are ultimately harmful to the environment. This includes direct emissions, such as those that result from driving to and from work as well as indirect emissions, for example the emissions from transporting the products we consume to the places we buy them.

Next, we will talk about the challenges rapid population growth pose to addressing climate change. Population growth essentially exacerbates problems related to climate change. More people consume more goods. Emissions and rates of deforestation rise more rapidly when birth rates are high – people must at the very least satisfy their basic needs for food, water, clothing and shelter, and this means the consumption of more resources.

Finally, we will talk about the regional distribution of the culprits of climate change. As you will see in Part II of this book, who is responsible for how much of climate change is a crucial part of climate science because of the politics of the issue. Thus, when talking about major culprits of climate change, we must consider who is responsible for what and how many emissions, as well as how a region's contribution to climate change might be affected by its economy and state of development. For example, in developing countries or countries with an agriculture-based economy, often one in the same, a substantial proportion of greenhouse gas emissions are methane and nitrous oxide emissions as a result of agricultural activities.[1] In more developed regions, in contrast, carbon dioxide from burning of fossil fuels accounts for more emissions than agriculture or deforestation. The ethical implications of regional perspectives on the distribution of responsibility for climate change are explored in Part II of this book.

Emissions from energy

Energy consumption is the leading source of greenhouse gas emissions.[2] According to the International Energy Agency, emissions from energy account for 64 per cent (almost two-thirds) of anthropogenic greenhouse gas emissions globally and 84 per cent of global CO_2 emissions can be traced back to the energy sector.[3] 'Energy emissions' refer to emissions from electricity production, transportation and heat supply. In the year 2000, energy levels were more than ten times higher than they were in 1900, which the IPCC attributed to higher rates of economic growth over the course of the century.[4] The World Energy Outlook for 2009 called for a 'wholesale transformation of the global energy system' in order to effectively address concerns related to climate change.[5] In order to limit the rise in global mean temperature to 2 degrees, says the International Energy Agency (IEA), levels of consumption from fossil fuel energy sources would have to peak in 2020 and decline thereafter.[6] It is not our energy consumption in and of itself that drives emissions up but rather fossil fuels as the predominant source of energy – over 80 per cent of our energy comes from fossil fuels. The burning of fossil fuels accounts for the largest share of anthropogenic greenhouse gas emissions.[7] According to the IPCC, 75 per cent of all carbon dioxide emissions from humans can be traced directly back to the burning of fossil fuels.[8]

So, 64 per cent of global greenhouse gas emissions come from fossil fuel industries. But this 64 per cent is by no means distributed equally across the globe. Regional distribution varies for all emissions sources, but the distribution of fossil fuel emissions from energy by country is an important factor in political and ethical debates. Objectively speaking, scientists say drastic reductions of fossil fuel emissions are necessary to climate change mitigation measures. However, such policies are controversial because the use of fossil fuels is associated with economic growth and industrialisation, and some countries are of the opinion that they would forgo critical development opportunities by eliminating fossil fuels as an energy source. Part II of this book details the ethical debates surrounding policies; therefore, it is important to have a rough idea of who is responsible for what when it comes to fossil fuel emissions in the energy sector. Industrialised countries and China are responsible for the lion's share of emissions from fossil fuel, both currently and historically. In 2000, the US and China alone accounted for 43.8 per cent of carbon dioxide emissions from fossil fuels.[9] As quickly developing countries like China, India and Russia join the ranks of top emitters, climate negotiations become more tense. Those 'new' to the fossil fuel industry claim they have the right to catch up to the established industrial powers, while these argue that top emitters are top emitters and the past is the past. We will discuss the merits and drawbacks of each of these arguments in Part II. After emissions from energy, deforestation and agriculture are responsible for the next largest proportion of anthropogenic emissions.

Deforestation and changes in use of land

The 16 per cent of carbon dioxide emissions that are not a result of burning fossil fuels are a product of deforestation and other alterations to natural landscapes and ecosystems.[10] Deforestation is a triple-hit to the climate. First of all, when trees are cut down they release the carbon they have thus far absorbed from the atmosphere is re-released back into it. Very old trees release the most carbon, because they have absorbed the most over time. Second, cutting down the trees eliminates an important sink for carbon dioxide, so no more carbon dioxide can be absorbed by those trees. Third, the land on which these trees grew does not sit idle. The land is developed for settlement, agriculture, or economic gain, all of which inevitably result in emissions. Thus, deforestation not only releases greenhouse gases into the air, it also eliminates an important sink for carbon dioxide and paves the way for additional emissions.

In the past, deforestation was associated with high or rising levels of development as it implied clearing land for industrial or residential use. However, due to improved understanding of the negative impacts of deforestation, today destruction of forests is considered an indicator of low development levels.[11] Deforestation occurring today is typically concentrated in developing countries. This is not to say that these are the only regions in which there is less forest coverage than in the past. The article 'Population and Environment' by Alex de Sherbinin points out that 'most developed countries largely deforested their lands in past centuries', thus leaving large forest areas in developing countries ripe for large-scale economic or agricultural development. The rapid population growth in these same regions results in additional deforestation for residential development, thus resulting in massive loss of forest area.[12] Figure 3.1 below from the human development report in 2011 shows the rates of deforestation or afforestation – that is, the re-planting of forests – in regions around the world.

The changes are most dramatic in Latin America and the Caribbean and in Sub-Saharan Africa. These regions also coincide with the regions with the highest birth rate. Land is developed for living space, sustenance farming and economic gain through trade. The ecosystems in these regions are fragile;[13] rapid changes to the environment therefore result in more extreme and dangerous changes than in other regions. As in many places, the natural landscape provides protection against natural disasters. The destruction of these natural barriers coupled with a higher frequency of severe storms and a lack of resources to invest in man-made protection, means inhabitants of these areas are most at risk to become

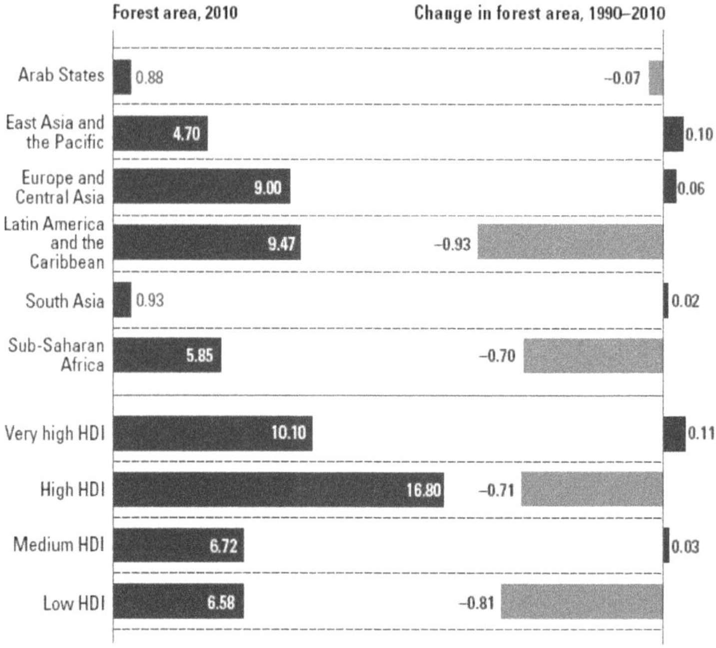

Figure 3.1 Forest cover shares and rates of change by region (1990–2010) (millions of square kilometres)

victims of the effects of climate change. De Sherbinin et al. write that after settlement, agriculture is the most common reason for clearing land.[14] Deforestation for the purposes of cultivation was once associated with development. Because of the environmental risks and destruction of resources that occur, though, it is now often more associated with underdevelopment.[15] Forests are also often cleared to make room for agriculture. This, too, occurs predominantly in developing countries, partly because industrialised countries tend to import agricultural goods from poorer countries where labour is cheaper and partly because there is simply more unexploited forest area in less developed countries. The next paragraphs will show how agricultural development further contributes to climate change.

Agriculture

Agricultural production is an indispensable part of sustenance, income and trade worldwide. One can see from the change in land cover alone the growing (no pun intended) importance of agriculture across the globe. In 1750, agriculture accounted for 6–7 per cent of land cover on earth.[16] By the 2007 IPCC Assessment Report 4, agriculture accounted for 40–50 per cent of total land cover.[17] Due to rising populations and higher rates of consumption, demand for agricultural products has increased.[18] In poorer countries, agriculture is a source of work for a majority of the women and helps keep many from falling victim to poverty and malnutrition.[19] With it, however, come serious consequences regarding climate change. Agriculture is said to cause between 10 and 12 per cent of emissions per year globally. Carbon dioxide emissions resulting from land cultivation are minimal with the exception of emissions from processing and transporting agricultural products, which are considered energy and transport emissions rather than agricultural. Most of the emissions from agriculture are methane and nitrous oxide.[20] Methane emissions from agriculture account for 47 per cent of total global anthropogenic methane emissions; it is emitted when organic materials decompose in an environment with no oxygen. In the case of agriculture, this comes from digestion in livestock, stored manure and rice farming.[21] The other major greenhouse gas emitted in agricultural processes is nitrous oxide. It is produced when nitrogen is transformed in soils and manures; nitrous oxide emissions in agriculture account for 58 per cent of global nitrous oxide emissions.[22] Emissions from agriculture are expected to increase as a result of higher food demands and changes in dietary habits.[23] We will talk about these changes in our discussion of rapid population growth as a factor of climate change below.

Rapid population growth

The world population has approximately doubled since the beginning of the 1960s and continues to grow. As can be seen in Figure 3.2 the 2012 world population is almost six times what it was in 1900 and most of that growth has take place since 1950.

Growing population plays a major role in rising levels of greenhouse gas emissions. More people mean more consumption inevitably resulting in higher emissions. More people require more land to live on and more food.[24] The problems presented by an increasing population are compounded by increasing rates of consumption. Not only are there more people, but the average individual is also consuming more. We mentioned this when talking about emissions from energy – rates of energy consumption have increased faster than population has grown due to economic development. Economic development leads to higher incomes and quality of life which in turn leads to more consumption, in this case more demand for energy. The higher population growth rates

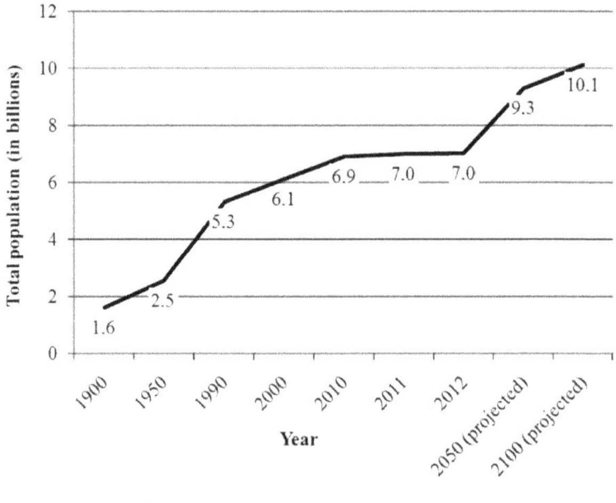

Figure 3.2 Global population growth in the twentieth century

are, the more difficult it is to keep emissions in check, even with measures to improve efficiency in emissions-related industries. In fact, in most places emissions intensity, or the amount emissions necessary to produce one unit of the GDP (gross domestic product), is falling but overall emissions are rising due to growing populations.[26] Industries are producing more efficiently, but there are more people consuming more goods.[27] The numbers are telling. From 1970 to 2007, emissions intensity, i.e., efficiency, improved by 40 per cent globally. Total emissions, however, rose 112 per cent and emissions per capita 17 per cent.[28]

The negative effects of rapidly growing populations are most conspicuous in poor and/or developing countries. Researchers and politicians alike are quick to remind industrialised countries that though their populations may be lower, the higher standard of living enjoyed by their citizens accounts for far more greenhouse gas emissions than emissions resulting from rapid population growth in developing countries.[29] They point out that the highest rates of population growth are typically in the least industrialised and poorest regions where the residents engage in fewer emission-generating activities than their wealthier counterparts in other regions. While historically true, this pattern is changing. China, for example, has one of the fastest growing populations in the world. Though still considered a developing country, China's economy is rapidly growing and rates of consumption are sky-rocketing, i.e., there are more people consuming more, which will result in an unprecedented rise in emissions levels. In fact, this trend is occurring worldwide. Later in the book, we talk about rising levels of development. For now, it suffices to say that global average levels of well-being as measured by the Human Development Index (HDI) are rising. As people's quality of life improves, so too do their levels of consumption. Indeed, this is one of the factors contributing to a rising quality of life as measured by gross domestic product (GDP). Consumption, as we will see throughout this chapter, necessarily means emissions. Thus, in many developing countries where both the population and HDI levels are rising, emissions are increasing that much faster. In the following pages, we will talk about the habits, activities and industries that result in the bulk of emissions.

Nutrition

Research has shown that historically, as a country's citizens' average income rises, its per capita emissions rate rises drastically as well before eventually levelling off.[30] Many poorer countries find themselves in this stage of development. The IPCC projects that future greenhouse gas emission rates will rise due to population growth and changing diets. People in developing countries are consuming more meat than in the past which means more emissions from livestock and agriculture.[31] In fact, the demand for meat in developing countries more than doubled between 1967 and 1997, from 11 kilograms of meat per person per year to 24 kilograms (from 24 to 53 pounds).[32] This trend is expected to continue through at least the first few decades of the twenty-first century and to occur mostly in South and Southeast Asia and Sub-Saharan Africa.[33]

A growing population consuming increasingly more meat poses a significant threat to the climate. The cost of meat to the environment is becoming more and more well-known. Nevertheless, the meat industry continues to flourish, to grow even, and at a rate disproportionately higher than the rate of global human population growth. The environmental damage occasioned by the meat industry is manifold. Why does meat 'cause' more emissions than other foods? For one, animal excretions, especially from cattle, release methane into the air.[34] If you remember from our breakdown of the agents causing climate change (last chapter), methane is more than 20 times worse than CO_2 for the atmosphere. Also, the animals have to be maintained until they are of the appropriate size and age to be slaughtered, which requires fertile land for grazing and sufficient food and water. Lastly, the meat must be properly stored and efficiently transported to prevent it from spoiling, and both refrigeration and transportation require significant energy, usually from fossil fuel industries.

Urbanisation and migration

Urbanisation is frequently discussed in connection with rapid population growth as severely exacerbating or accelerating anthropogenic climate change. The urbanisation of coastal cities, especially, which is due in part to rapid population growth, creates further vulnerabilities in an already climatically threatened environment. In the last chapter, we talked about rising sea level and how this endangers coastal regions and the species that inhabit them. Depletion of the natural resources in these regions due to urbanisation further weakens the ecosystem and natural defences against hurricanes, tropical cyclones and the like.[35] De Sherbinin et al. point out that 14 of the 17 largest cities in the world are on a coastline, and 'coastal and marine environments continue to be among the most threatened ecosystems in the world, owing in part to the sheer scale of detrimental human activities associated with urbanization along the coasts, continued population growth, and a growing number of tourists in search of coastal amenities.'[36]

There are, however, other effects of large concentrations of people in cities that apply to coastal and land-locked cities alike. Urban populations consume between 60 and 80 per cent of total global energy,[37] and if you remember from above, the energy sector is responsible for 64 per cent of global greenhouse gas emissions. As with changing dietary habits, the effects of urbanisation will be most conspicuous in developing countries.[38] Developing countries, however, are still industrialising. As they continue to develop, more people will be drawn to cities for economic opportunities. Emissions will rise for two reasons: first, the industrialisation that provides job opportunities also produces emissions. Second, as income rises as a result of new jobs, consumption grows as well, which means even more emissions.[39]

Migration patterns associated with rapid population growth and demographic changes also pose a formidable threat to the climate. Since the beginning of civilisation, humans have sought out settlements near sources of water for access to nourishment and transportation as well as for trade possibilities and protection from enemies. Today, coastal areas continue to attract multitudes, mostly for the potential from economic gain in the tourist industry or from as yet unexploited natural resources and/or additional available living space required by a growing population.[40] Natural coastal landscapes are being converted to seafood farms to feed the growing demand for seafood, thus damaging natural coastal protection and taking away natural habitat of many fish species.[41]

The tourism industry is also proving to be a stress on the environment in coastal regions. We talked about the disappearance of coral reefs and the consequences for the impact of tsunamis in affected areas. Rapid population growth and the expanding tourist industry exacerbate the negative effects climate change is having on coral species. As the number of tourists visiting tropical areas increases, the demand for coral as a souvenir increases. Local residents from nearby overcrowded cities or rural areas then migrate to the coast to capitalise on the rapidly evolving tourist industry. This puts a strain not only on the natural resources sold as souvenirs, but also on the land developed for residences, hotels, shops, restaurants, etc., and the fragile ecosystems that serve as a natural protection for coastal lands. Further deteriorating the quality of the natural resources is the pollution that inevitably comes with such development and puts significant strain on the environment.[42] Such trends are all the more dismaying when one considers that tourism *could* have positive effects on an environment. Profits from the tourism industry could be used to invest in infrastructure that could protect the area from climate change or to invest in technology that could reduce emissions and thus improve the environment. Also, tourism can be used to call attention to the beauty and resources of different regions and landscapes as well as raise awareness for the threats posed by climate change to local inhabitants. Unfortunately, the areas in which tourism is used so constructively comprise a small minority of 'destination' locations.

4

THE HUMAN COSTS OF CLIMATE CHANGE

We have covered how, why and how much Earth's climate is being affected by human activities. Now it is time to talk about the consequences of these changes for our lives and lifestyles as well as those to come after us. These are the most difficult costs to quantify in climate science. Though there are formulas for measuring a person's quality of life or well-being, there is no variable or room for inserting climate change into this formula. Disease and death rates, quality of life, levels of poverty and other aspects to be considered in calculating levels of well-being build a complex web of factors that make it impossible to say with certainty the cost of climate change to human livelihoods and well-being.

However, we can say how various ecosystems and environments would be affected by changes in climate. We can also say how many people rely on or live in affected systems or regions. From there we can deduce what consequences what people will suffer at the hands of climate change.

As is often the case, poorer people and individuals will be disproportionately affected by the negative consequences of climate change. The 2011 Human Development Report emphasises the reliance of the rural poor on natural resources both for sustenance and income. Fourty-Five million people who rely on fishing for food and income will be affected by the expected significant decline in fish populations in the Pacific Islands. Three hundred fifty million people living in or near forests rely on their resources for subsistence and income. Many of these may end up with in a Catch-22. Deforestation results in the destruction of the resources on which they rely, and as governments begin to try to preserve remaining forest cover, those dependent on these resources may be faced with restricted access that prevents them from meeting their needs. In all of these instances, women are more affected than men. More women are dependent on these natural resources because of limited employment opportunities for females in developing countries or regions. The loss of land and forests and the degradation of hitherto agriculturally fruitful terrain do not just affect people in developing countries whose livelihoods are dependent on these resources. A grand total of 1.3 billion people involved in agriculture, fishing, forestry, hunting and gathering are expected to be directly affected by changes in landscape due to climate change as well as the ways people cope with it. Additionally, food prices are expected to rise dramatically – by as much as between 35 and 50 per cent – as fertile land areas degrade, and this affects all of us.

As a result of disappearing natural resources and higher food prices, many experts are prognosticating the onset of or increase in what they have dubbed 'climate wars'. Climate wars would refer to civil unrest and conflict due to insufficient food, shelter and income arising from climate change thus leading to uprisings, protests, and ultimately war. The IPCC acknowledges this possibility but reminds readers that isolating climate as a deciding factor in such complicated conflicts is difficult. Some speculate as to whether the recent conflict in Darfur was the first of the climate wars.

Speculations have also been made as to potential ethnic conflicts when people are forced to migrate due to barren lands or inundation from rising sea level.

Even in the absence of civil or ethnic conflicts as a result of climate change, rising temperatures pose both direct and indirect threats to the well-being of people all over the world. Rising temperatures make heat waves much more likely. Some researchers have concluded that the likelihood of the heat wave in Europe in 2003 resulting in the death of tens of thousands of people was considerably increased by the rise in global mean temperature by 0.75°C since pre-industrial times. Such heat waves in developing countries would prove to be even more catastrophic, as fewer of those affected would have resources such as air conditioning or other cooling systems to cope with the heat. Additionally, rising temperatures affect crop cycles. In northern latitudes, this would mean longer growing seasons. However in tropical climates, this would mean longer dry periods and damaged crops. Harvests in Sub-Saharan Africa would also be affected. In other words, those who can least afford it would take the hardest hit.

The caricature in Figure 4.1 below is worth a silent snicker or a wry smile at the least. Unfortunately, though, it is not an exaggeration of absurd proportions. In fact, only the talking sharks make it unrealistic at all. In Chapter 2 we talked about the potential of rising sea level to cause permanent flooding covering a surface area of the Earth equal to the surface area of Italy and France combined. The consequences for coastal populations are nothing short of catastrophic. Although only 2.2 per cent of Earth's land cover is less than 10 metres above sea level, this 2.2 per cent is inhabited by 10 per cent of global population; this means well-above average population densities in these areas. This means that should a significant piece of this surface area be

Figure 4.1 Caricature illustrating global warming

flooded, large numbers of people would be displaced and forced to migrate elsewhere. For those areas not completely inundated, increased flooding will prove to be a formidable challenge and a threat to the lives and well-being of those affected.

These are not the only costs associated with higher water levels and more flooding, though. If sea level continues to rise, entire cities could be erased from the world. According to the 2011 Human Development Report, more than 63 million people in East Asia and the Pacific regions alone will be affected. Even more people will be affected in countries with a high HDI, but poor regions will incur the most damages because they do not have the resources to invest in infrastructure that would protect them from the dangers of severe flooding. Rapidly rising number of extreme weather events will exacerbate damage from flooding and are likely to result in an increase in levels of disease and death that can be traced back to environmental factors. The costs will differ across various demographics and not just based on income. Women, children and the elderly (i.e., the traditionally weaker members of society) especially in poorer regions face a higher risk of injury or death from high winds and floods.

As you can see, climate change is not some distant, abstract concept. Our everyday lives are and will be affected by global warming. In turn, almost everything we do on an everyday basis contributes in some way to mitigating or exacerbating climate change, even if on a miniscule scale. We must act not only for ourselves but also for future generations. We will see in Part II that this sentiment is not simply a gesture of charity but rather a moral imperative. Before we show why even the most basic code of ethics demand that we act, we will discuss how we can address climate change and the challenges we face when doing so.

5

ADDRESSING CLIMATE CHANGE: OPTIONS AND OBSTACLES

Since the dawn of international climate negotiations, countries have been reluctant to implement measures to mitigate or adapt to climate change because of the economic repercussions. No one is willing to forgo economic growth for the sake of the environment without a guarantee of sorts that other parties will make the same sacrifice. Heads of state and politicians fear that reducing emissions would mean abstaining from the economically profitable activities from which these emissions result. In the media, economic growth and climate change adaptation/mitigation are depicted as mutually exclusive objectives between which governments and citizens must choose. Some leaders attempt to rationalise not acting on climate change by saying the future will be able to adapt adequately with still to be invented technologies or other mitigation and adaptation measures. Even if significant progress is made in climate change mitigation and adaptation, there is no way to know when or how. As we will see in Part II, this stance is not ethically acceptable. More to the point, there is no reason to wait for the future when there are options for addressing climate change here and now.

Proposals for mitigating or adapting to climate change run the gamut from painting rooftops white so that they reflect solar heat waves, to burying carbon dioxide underground, to renewable energy instead of fossil fuel based energy and everything in between. We will confine our discussion of options for addressing climate change to those measures relevant to our climate ethics discussion in Part II. This essentially means measures that can be implemented at the national and international levels because players at more micro levels can typically not significantly affect major climate policy. We will cover two major climate policy areas – energy policy and population policy. The energy sector holds the most potential for reducing emissions both because it is the largest source of anthropogenic emissions and because of the many options available for improving efficiency or for alternative energy sources altogether. Population policy is important because of its controversial nature. It is essential that population growth rates fall if we want to get a handle on climate change; however, one can also easily wander into murky ethical territory when talking about measures that would influence very personal family matters.

Efficiency and alternative energy

Energy represents a gargantuan portion of global emissions and is thus one of the most important areas in terms of emissions reductions. While environmentally friendly alternatives to the currently prevalent fossil fuels used for producing energy exist, these alternatives are often depicted as 'luxury' items – they are expensive, sometimes unreliable and one must live in a geographical region well-positioned to generate enough renewable energy to meet demand. Furthermore, there is no single technology that could reduce emissions or replace our demand for fossil fuels. However, developing the market for renewable energy (solar, wind and water) could offer an attractive alternative to fossil fuel energy. In its special report on renewable resources, the IPCC states:

> As well as having a large potential to mitigate climate change, [renewable energy] can provide wider benefits. [Renewable energy] may, if implemented properly, contribute to social and economic development, energy access, a secure energy supply, and reducing negative impacts on the environment and health.

Perhaps the best way to reduce energy emissions both in the short- and long-term is to improve the efficiency of our energy use. By making the most out of our vehicles, appliances, homes and industries, we make the most of the energy we have, thus reducing our energy needs and emissions levels.

Government policies can prove a very effective instrument in transforming the global energy market. The surest way to ensure fewer emissions in the energy sector is through taxes and subsidies. Government subsidies for renewable energy would encourage research and development as well as market investments in clean energy. On the other side of the spectrum, taxes on fossil fuels are needed that reflect the true costs of these industries, i.e., the cost of emissions of fossil fuels to the environment. On the demand side we as consumers need to adjust our consumption to more environmentally sustainable levels.

Addressing population growth

There is no getting around it: in order to adequately address climate change, the problem of rapid population growth must be addressed. Many people are sceptical of, or outright opposed to, government policies aimed at influencing national population growth, thus necessarily influencing reproduction practices in the country. Opponents of all forms of population policy argue that reproduction is an area that is strictly personal in nature and should be off-limits to government interventions or influences. Measures such as the one-child-policy in China have stigmatised population policies in general, but there are morally acceptable ways of encouraging reductions in population growth rates without infringing on personal freedoms. In an article on global demographic trends, Brian O'Neill et al. point out that

> household surveys indicate that there is a substantial unmet need for family planning and reproductive health services in many countries. Policies that meet this need would reduce current fertility by about 0.2 births per woman in the United States and 0.6–0.7 births per woman in the developing world.

According to these authors, reducing global population growth rates is absolutely necessary in reducing emissions, and about one-half of these emissions reductions could be achieved by implementing such policies in the United Sates and developing countries (with the exception of China).

Family planning programs are considered population policy and are quite the opposite of an infringement on personal freedom. Indeed, such programs empower women and give them the information they need to make their own decisions about the size of their families. Making contraceptives available in developing regions where access to such protection has been limited in the past is also an important and simple measure to 'influence' population growth rates. The 2011 Human Development Report cites Bangladesh as a country with successful population policy strategies – between 1975 and 2009 birth rates went from 6.6 births per woman to 2.4

births per woman. The change is attributed to government efforts to make contraception more widely available as well as political conversations with leaders of both genders and similar campaigns that transformed the perspective of the general population and influenced social norms.

Cooperation between governments, social scientists and natural scientists is necessary to effectively reduce population growth rates in an ethically acceptable manner. Here, we again come across the problem of interdisciplinarity. In an article on population and environment, de Sherbinin and others show how, until recently, researchers in the natural sciences have tended to oversimplify population-environment interactions. '[P]opulation can be treated in models in a manner that is analogous to all the other quantitative variables', they say, '[t]his has promoted something of a reductionist view of population-environment interactions.' This is beginning to change, and it is important that it does. Those in the natural sciences must be able to understand the language of social sciences in which absolutes are rare since people cannot be 'reduced to food and material demands that result in some aggregate impact on the environment'. Values, cultures, institutions and social norms must also be taken into account when considering population impacts on the environment and options for reducing these impacts. Further challenges to implementing climate policy are discussed in the next chapter before we move on to ethical aspects of the climate change conundrum.

In Chapter 1 we talked about some of the most formidable challenges to bringing climate change to the forefront of public attention and to putting pressure on policymakers to implement climate policy. In addition to these factors, the economic implications of stringent climate laws make the general population and international leaders alike hesitant to take a progressive approach in addressing climate change. The discourse on how to monetise the damages to the environment and climate and the substances that cause them is a natural extension of the discussion of anthropogenic climate change and the dangers that accompany it. For one, activities and industries that produce the emissions responsible for an abnormally rapid climate change are economic enterprises and thus naturally driven by the goal to turn as high a profit as possible. Incorporating the relatively newly determined costs of climate change into the well-established market mechanisms that determine prices and ultimately profits presents a unique challenge to economists. We won't go into the details of the numerous theories and proposals for putting a monetary value on environmental resources and calculating the costs of emissions in terms of dollars. The economics of climate change is a genre of literature all its own. However, it is important to present a brief overview of the (value of) resources at stake because of the role the economy plays in politics and furthermore its status as a 'competitor' to ethical arguments.

The ultimate goal of any business venture is to turn a profit. Pollution is typically not considered a cost of production. In order to curb emissions to the extent necessary to abate climate change, the true costs of emissions have to be calculated as production costs no different from the costs of material, labour, etc. Some companies have to consider the costs emissions certificates they have to buy, but according to most economists and environmentalists, the costs of these certificates are not indicative of the true costs of the emissions in terms of damage to the environment.[1] In an article from 2009, Muller et al. calculated that four major industries – stone quarrying, solid waste combustion, sewage treatment plants and fossil fuel based power generation – actually have pollution costs that are higher than profits as measured according to traditional factors.[2] In other words, if companies in these four industries had to pay the actual costs of their emissions, they would no longer be making a profit. This does not suggest that these industries should be shut down. Rather, it exemplifies the importance of mechanisms that accurately measure the costs of emissions in order to effectively and efficiently address climate change issues. 'It is instructive', the authors conclude, 'that many of the current efforts at climate-change mitigation are aimed at automobiles, while the largest damages are likely to be in the area of coal-fired generation'.[3]

In 1997, a group of environmental scientists and economists wrote an article concerning the value of Earth's natural resources. According to their calculations, the Earth's natural resources are worth between US$ 16 and $ 54 trillion, with an average of $33 trillion per year. There are admittedly a lot of uncertainties in such an equation. The authors addressed this, saying, '[b]ecause of the nature of the uncertainties, this much be considered a minimum estimate. Global gross national product total is around US$ 18 trillion per year'.[4] That means that even at the lowest of the minimum estimates, our natural resources are worth as much as, and probably more than, the entire world's sum of products and services. Though too technical for our purposes, solid, established economic theory and methods back up the article's conclusions. However, until the actual value of the natural resources and stake and the actual costs of anthropogenic emissions are integrated into classic cost-benefit analyses, the economic costs of stringent climate policy will appear to outweigh its benefits.

Exacerbating the problem are powerful, wealthy interest groups and corporations constantly lobbying politicians and public opinion in their favour. For those who stand to lose should stringent environmental regulations be implemented and/or enforced, climate science lends itself nicely to different interpretations. In the previous four chapters, we discussed the science of climate change, potential consequences and potential measures for adaptation and mitigation. Those who have a stake in challenging the passage of climate policy cherry-pick to build a seemingly plausible, science-based case against the reality of climate change. In addition, these interested parties have launched extensive propaganda campaigns in support of their views. First we will review the arguments brought forth by climate sceptics, as they are known, and point out the flaws before analysing the propaganda campaigns put forth by sceptics and others who stand to lose should strict climate policy be implemented.

Scientifically disproving climate change?

When the first IPCC report was published in 1990, many parties challenged the existence of global warming. Today, not even the staunchest climate sceptic can deny global warming or climate change. Though there are many uncertainties in climate science, there is absolutely no doubt that there is indeed a natural greenhouse effect without which our planet would be (and was at one point) uninhabitable. Still, even today there are initiatives such as the International Climate and Environmental Change Assessment Project (ICECAP) that attempt to play down climate change.[5] Climate change sceptics often emphasise that the greenhouse effect is a natural phenomenon without which Earth would not be a hospitable environment for people. Furthermore, they say, any effect that anthropogenic emissions may be having on the climate is miniscule in comparison to the natural greenhouse effect. All of this is true and, taken out of context, could be taken as evidence that warnings about potential dangers of anthropogenic climate change are simply alarmist and overreactions. There is indeed a natural climate change. Without it, the average temperature on Earth would be –18°C (–0.4°F) and in most places inhospitable to human life.[6] In comparison, a paltry 0.75°C warmer hardly seems significant. As we made clear in Chapter 2, however, the Earth's ecosystem exists in a fragile equilibrium that developed over thousands of years. Thus, a 0.75°C increase in the relatively short time span of a few centuries is most certainly significant.

What is more, the natural greenhouse effect and the anthropogenic greenhouse effect differ in how they affect the climate or, more specifically, in what substances contribute how much to warming the planet. Water vapour contributes the most to the natural greenhouse effect, with carbon dioxide coming in at a distant second. In contrast, we know from Chapter 2 that carbon dioxide plays by far the biggest role in the anthropogenic greenhouse effect, followed by methane; water vapour contributes a small, almost insignificant forcing in anthropogenic global warming. Furthermore, there is no doubt as to *what* excess carbon dioxide in the atmosphere is due to anthropogenic emissions, because naturally occurring CO_2 atoms differ in weight from those resulting from human activities.[7]

Many climate sceptics use individual weather events that seem to contradict the concept of global warming as proof that climate change is at best an exaggeration. At the same time, they use the climate experts' statements that individual weather events cannot be predicted by or attributed to climate change as proof that climate science is simply a theoretical guessing game.[8] Furthermore, climate sceptics purport that the methods of measurement of the IPCC and other prominent climate organisations are inadequate and inaccurate. According to these assertions, measurements from stations around the world are only representative of the region in which the measurement was taken and even these can be faulty due to technical malfunctions or human error in calculations.[9] This suggestion is more a display of ignorance than anything else, because measuring tools and models have long since taken regional differences into account in their calculations.[10] Today the data are solid and methods are constantly being refined and improved. In addition to publications and websites, climate sceptics have launched anti-global-warming propaganda campaigns in the popular media, partly financed by industry giants with a vested economic interest in preventing the passage of climate laws (the transportation industry, the energy sector, oil companies, etc.). The next and final section of Part I will discuss some of the highlights of these campaigns and their repercussions.

Media in the middle tug-of-war: climate vs. capitalism?

Many who have observed the progression of the depiction and status of climate change in the media have drawn parallels with that of cigarettes and the tobacco industry.[11] Just as tobacco giants played dumb for decades about the hazards that smoking poses to one's health, top executives and researchers in the oil industry played down or denied (the effects of) climate change for decades until political pressure from the scientific community forced them to at least acknowledge climate change. The advertisement headline below, from the since disbanded Informed Citizens for the Environment, is one of the climate sceptic advertisements widely circulated early on in the climate change debate. Not only does the advertisement compare global warming to the centuries-old belief now recognised as ludicrous that the Earth is flat but it also implies that global warming is a foolish belief of the masses, and that only the educated, enlightened few know the truth, and

Figure 5.1 Advertisement headline of the organisation 'Informed Citizens for the Environment'

that climate change 'believers' are simply members of the masses who accept what supposed experts tell them at face value. Other propaganda campaigns that we discuss below were more subtle, but the message is always the same:

Don't be fooled by alarmists overreacting to small or non-existent phenomena.

ExxonMobil Corporation's contribution to what is now considered a disinformation campaign is probably the most conspicuous. Exxon was one of the last of the major oil corporations to admit that there was indeed undoubtedly anthropogenic climate change and emissions from fossil fuel burning contributed significantly to the growing crisis. In its days of climate denial, Exxon was part of, and provided funding for, organisations with names like the Global Climate Coalition which worked to counter the work of organisations and scientists like the IPCC.

For example in December of 2001, an article on the Global Climate Coalition's website stated that the warming of the earth resulted for some regions in longer growing seasons and could thus ultimately result in a higher rate of CO_2 absorption by plants. The same article suggested that because scientists were uncertain as to the annual exchange rate of carbon dioxide in nature, i.e., between land, air and water, they could not be sure if anthropogenic emissions were not actually 'canceled out' by natural processes.[12] This is one example of some 70 articles on the Global Climate Coalition's website just for 2000 and 2001, many of which imply or state outright that the effects of global warming were uncertain and debated. Regardless of whether these stories were written out of an obligation to present both sides of the story or to depict climate change as uncertain and controversial the effect was the same: an impression of uncertainty and confusion among experts in the field of climate science. Understandably, no one wants to support measures, policies or laws that are expensive and inconvenient and based on scientific results that could, by all appearances, be overturned at any moment.

Though eventually forced by overwhelming evidence to concede to the existence of climate change, as late as 2005 Exxon purported in their Corporate Citizenship Report that 'the extent to which recent temperature changes can be attributed to greenhouse gas increases remains uncertain'.[13] A severely critical and widely publicised letter from Nick Thomas, the director of public affairs of the Royal Society, Britain's premier scientific academy, chastised the company for this and similar statements.[14] Exxon refuted the charges and claimed that '[t]he Royal Society's letter and public statements to the media inaccurately and unfairly described our company'.[15] Despite this cry of foul play, however, subsequent Corporate Citizenship Reports refrained from any suggestion that climate change may not be real, settling instead for the assessment that 'the risk to society and ecosystems from rising greenhouse gas emissions could prove to be significant'.[16]

Without a doubt, the propaganda campaigns were/are a huge setback for climate change prevention and mitigation measures. Campaigning was continued by dropping back to the periphery and taking the more low-profile route by financially backing 'independent' scientists.[17] By 2008, Exxon had pledged to cease its financial support of organisations publishing or backing disinformation regarding climate change.[18] However, a wave of media reports in 2009 and 2010 called the corporation out for not following through on this promise. 'Exxon. . .gave reassurances [in 2009] that it had no funding links with the sceptics' biggest annual conference, the International Conference on Climate Change. But a list published by Exxon. . .of its '2009 worldwide contributions and investments' revealed it had given four co-sponsors of the New York event $US275,000. It also gave $US1m to 20 other sceptic groups.'[19] Finally, in 2011,

Exxon made good on their word. That is to say, in 2010 ExxonMobil contributed less than $800,000 to organisations and individuals critical of climate science.[20] Compared to its all-time high of $3.4 million, this is indeed quite a drop. Since then, Exxon has also invested millions in research on making fuel out of algae,[21] but the company is far from being redeemed in the eyes of environmentalists and climate scientists. The research director for Greenpeace USA was quoted in the *New York Times* as saying Exxon's declining role in the furtherance of climate denial was not enough in and of itself. Indeed, he told *The Times*, 'It is like leaving a room without ever saying they are sorry'.[22]

Unfortunately, the minimalised role of oil and energy giants in campaigns regarding climate science does not by a long shot mean the end of the climate denial industry. For one, the fossil fuel industry's funding of those seeking to cast doubt on the credibility of legitimate climate science has only been significantly *reduced*, not eliminated. Secondly, we only mention here a *few* of the ways corporations and big business *directly* influence discussions concerning climate change. Such industries can afford to encourage lawmakers to see their side of things by donating to election campaigns and hiring lobbyists to champion their cause to national law-making bodies. There is nothing illegal about this. Corporations are not concerned, indeed are happy, when their interests weigh unfairly in lawmakers' decisions.

By now, the mainstream debate no longer revolves around whether climate change is real, but rather what the magnitude of climate change is and what to do about it. Natural scientists continue to refine and improve techniques and technology for measuring the effects of climate change. Politicians and international leaders continue to argue about who is at fault and who should shoulder the costs. Meanwhile, anthropogenic, or people-made, changes continue to create imbalances in ecosystems. Emissions are rising when they should be falling and this has negative effects on the earth and humanity. As we will see in Part II, ethical considerations demand that we act swiftly. Now that we have a basic understanding of the practical issues surrounding climate change we will delve into the moral factors that play a role in trying to build an international consensus on climate policy.

PART II

CLIMATE ETHICS

On December 19, 2010, the last day of the global climate conference in Copenhagen, several countries with high greenhouse gas emission levels proposed a painstakingly negotiated compromise paper to the plenary assembly. The chief negotiator of the developing countries, the Sudanese Lumumba di-Aping, called this 'a solution based on values that funnelled six million people in Europe into furnaces'.[1] This comparison to the Holocaust caused a worldwide sensation. One cannot reach more deeply into the console of moral arguments. But is such a comparison justified? This is a question neither for natural climate science, nor for the economics of climate change, nor for empirical political science, but rather singularly suited for ethics. The new and developing field of 'climate ethics' applies standard ethics methods[2] to climate change. Within this field, moral philosophy discusses how an individual person should act with regard to climate change. Political philosophy debates distributional regimes for greenhouse gas emissions and who owes what to whom. Articles on climate change are playing an increasingly important role in political philosophy but the number of ethical articles on the matter is still significantly lower than one would expect given its importance. Gardiner ascribes this to the unavoidable interdisciplinarity of the topic, which we talked about in Chapter 1 of Part I.[3]

What is fair when it comes to climate issues?

In this half of the book, we will examine what is 'morally right' in dealings with the climate. Normative arguments are employed frequently by interagents on all sides in climate negotiations. Regardless of ethicists' evaluations, politicians will continue to use pseudo-moral arguments in the future, if for nothing else than to score points with the media and civil society. To what degree the politicians are actually willing to give moral ideas precedence over national interests is another question (and will not be explored here). Our goal in this section is to use methods accepted in the field of ethics to differentiate the 'morally right' from the 'bad'.

Gardiner calls the ethics of climate change 'a perfect moral storm',[4] emphasising the especially intricate nature of the problem of climate change for political philosophy, a Pandora's box involving questions of pure distributive justice, international justice, historical justice, intergenerational justice and compensatory justice. There is no dearth of literature on climate ethics breaking down the complexity of the issue, thereby enabling one to arrive at partial conclusions such as: 'Historical Justice demands we do this...' or, 'intergenerational justice demands we do that...'. The following chapters, in contrast, attempts to face up to this complexity, that is: to end with a synthesis of the arguments into what I consider to be the fairest approach to the politics of climate change on a global scale.

However, before addressing the question of *justice* at all, we must first clarify what good or burden is to be distributed.

6

DISTRIBUTION OF WHAT?

In the field of climate ethics, some have presented convincing arguments for regarding the climate system as a 'global common'.[1] The atmosphere is possessed by no particular people, groups or countries. Given this lack of ownership, every human being is entitled to use of the atmosphere. In other words, each person has the right to emit a certain amount of greenhouse gases (carbon dioxide equivalents = CO_2e) into the atmosphere. The capacity of the atmosphere to absorb these substances, however, is limited. The demand for atmospheric resources exceeds the supply. From a moral point of view, it makes no difference whether the commodity in question is a resource or a sink. Principles of distributive justice among individuals can be applied just as well to emission distribution rights as to the sharing of other scarce goods (food, water, living space).

Prima facie the good to be distributed (in other words: the currency of climate ethics) is the capacity of the atmosphere to absorb greenhouse gases. But on an axiological level, one could imagine alternatives, for example a) the well-being produced by greenhouse gas emissions or b) the economic growth they generate or c) the costs of mitigation and adaptation.[2] The former is at the very least atypical terminology.

If we are talking about the just distribution of a load of rice in a refugee camp, then on an axiological level, we normally regard the rice as the distribution object, not the changes in well-being it brings about for its recipients. We usually calculate in sacks or kilograms, because we cannot know how much personal well-being comes from a sack of rice. Of course, we distribute the rice because we are concerned about the well-being (or even survival) of the refugees, but we count the ration that everyone receives in a physical measurement. This is so because incremental changes in most goods, and certainly emission budgets, cannot be directly translated into changes in human individual well-being. On a personal level, imagine you had the choice between an emission budget and a compensatory sum of money from an emission trading scheme. All of us would pick the money – precisely because an emission budget has *no* direct connection to well-being.

On a country level, human well-being can be measured, for example, by the Human Wellbeing Index, the Weighted Index of Social Progress, or the Human Development Index (HDI), which is the most accepted indicator of human well-being.[3] The ranking of countries with the highest current greenhouse gas emissions differs significantly from a ranking of countries according to their citizens' well-being, as Figure 6.1 shows. It is thus inadequate to replace the absorptive capacity of the atmosphere (one of its functions, measured with a physical metric) by the well-being we draw from it.

Option b), that is, to view 'economic growth' or 'chances for development' as the object of distribution, is even more problematic. As demonstrated below, the relationship between greenhouse gas emissions and chances for development are complicated. Economic growth does not result from greenhouse gas emissions alone; other ingredients

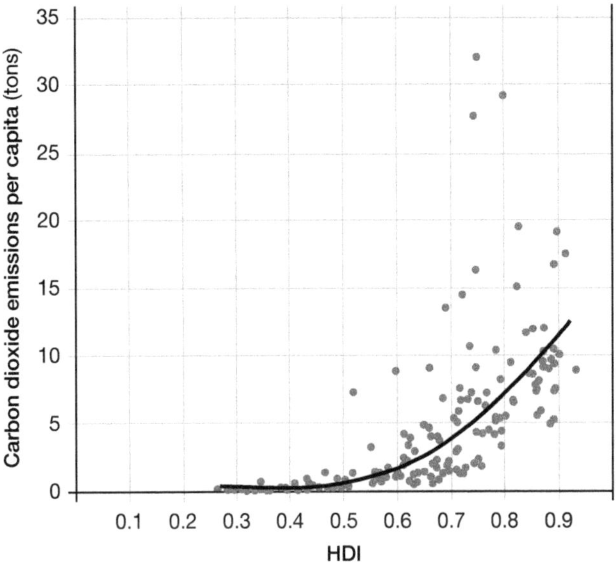

Figure 6.1 Relationship between quality of life (HDI) and current CO_2-emissions per head

(hard work, for example) are necessary as well. And, most importantly, a high gross domestic product (GDP) has no intrinsic value in philosophy; rather, it is a means to an end. In welfarism, this end would be human well-being.

A further suggestion is to not distribute greenhouse gas emissions but rather to distribute the costs of mitigation as measured by lost GDP-growth.[4] Miller calls this 'the principle of equal sacrifice.' This would mean that countries in which reducing greenhouse gas emissions could only be achieved at a high cost (hard-to-reach fruits) would conserve, say, 10 tons of CO_2e by investing 1 per cent of their GDP. Countries that have done little for climate protection up to this point and still have numerous easy possibilities for conservation (low-hanging fruits) would save 100 tons of CO_2e with a one per cent investment of their GDP. Despite some positive secondary effects, this proposition must also be repudiated. First of all, Millers premise 'that the reductions required will in almost all cases be costly in terms of foregone economic growth and personal consumption',[5] is disputable because it ignores population change as a major factor in climate change. As long as couples in many parts of the world are forced to have more children than they want to have due to a lack of contraceptives, one cannot classify 'having less children' as 'costs' (be it direct costs, opportunity costs or symbolic costs such as personal sacrifices). Secondly, many studies in happiness research have shown that an increase in personal income does not lead to more happiness once a threshold is passed.[6] As the major per head emitting countries (especially the United States) are way above the threshold that decouples increases in income with increases in happiness, the 'principle of equal sacrifice' would in fact mean that these countries would have to sacrifice GDP-growth, but not happiness. In contrast, poorer countries would have to make a sacrifice both in GDP-growth and in happiness.

Thirdly, how high the GDP in a particular country during a given reporting period *would have been* without measures of climate protection is hardly legitimately quantifiable (as Miller admits). It is even possible that such measures *increase* the GDP, because some goods (e.g., CO_2-capture complexes) and services (e.g., advising the population with respect to building insulation) would be included in the GDP calculation.

After scrutinising these options we conclude: *greenhouse gases* or its correlative, the absorptive capacity of the atmosphere[7] should be regarded as the climatically relevant object of distribution. Of course, they are no synonyms, but in the context of the object of distribution they are interchangeable, as one is the function of the sink, the other one is the thing that fills the sink.

The following chain of reasoning applies:

1) Human and environmental suffering as a result of climate change should be avoided. Consequently:
2) A global temperature rise, a rise in sea level and an increase in extreme weather events must be minimised. Consequently:
3) The cumulative amount of greenhouse gases in the atmosphere must be stabilised in the long term.

To stabilise the accumulated amount of greenhouse gases in the atmosphere in the long term, the global emissions budget for greenhouse gases between 2010 and 2050 is 560 billion tons of CO_2e (560 Gt CO_2e).[8]

7

INTERGENERATIONAL JUSTICE

How should atmospheric resources be divided between the members of present and future generations? As stated above, presently living generations should not fill up the atmosphere with more than 560 billion tons of CO_2e between 2010 and 2050 if a dangerous rise in global temperature is to be avoided. This notion of a safe emissions budget is a restriction for the current generations' consumption of atmospheric resources during their lifetimes. Theoretically, we could arbitrarily set this available budget to 1,000 billion tons and gain more leeway for distribution schemes in the present, e.g., between the North and the South. This manoeuvre – beneficial for the existing generations, but at the expense of future generations – would contradict what intergenerational justice demands from us. Our moral obligation to stick to the safe emission budget[1] and thus to prevent dangerous climate change, rests on two arguments: the potentially cataclysmal consequences of climate change, and the low costs of a second-order error.

The problem of *uncertainty* is crucial in this puzzle.[2] Each generation faces the problem that they can only vaguely predict how effectively and efficiently the next generation could reduce emissions and adapt to changing climatic conditions.

According to Karl Popper, we are fundamentally unable to make (long-term) predictions, because scientific and technical progress cannot be foretold. Indeed: a reliable, serious statement about the evolution of the quality of life of future mankind is neither on a global nor on a regional scale possible. The time-delayed nature of the consequences of our climate-related actions compounds the problem. However, there is strong evidence that the greenhouse effect may occasion many additional deaths and enormous costs for coming generations.[3] Of particular relevance are the so-called tipping points of climate change. This refers to changes that begin when a certain concentration of greenhouse gases in the atmosphere is reached and, once in motion, are self-accelerating and irreversible. The climate system is a non-deterministic system with an almost infinite number of variables, in which processes occur according to the chaos theory. In such a system, one cannot predict with certainty what will happen, but rather only probabilistically with a higher or lower degree of certitude.[4] One of the most feared consequences of human-induced temperature rise is the melting of the Greenland ice sheet. This would cause the sea level to rise 6–8 metres; several countries would completely disappear from the map and others would permanently lose large coastal cities. No serious scholar can eliminate this possibility or definitively prognosticate its occurrence. Moreover, an objective probability is unquantifiable and to provide a subjective probability based on current data seems speculative. From an ethical standpoint, mid-level magnitude damage with a medial occurrence probability has to be judged differently than highly destructive damage with a small probability of occurrence, even if both scenarios produce the same expected value. To sum up, the expected value principle is suitable for the medium range, the precautionary principle for the extreme.

Another argument for the precautionary principle is the low cost of a second-order error. In other words: if we remain idle because we assume that there is no man-made climate effect/temperature increase/rise in sea level and we are wrong, the consequences are catastrophic. If we change our behaviour because we believe that there is a man-made climate effect/temperature increase/rise in sea level and we are wrong, the consequences are not dramatic in the long run. Fossil fuels will be exhausted in the foreseeable future; therefore, it does not make much difference whether we expedite the transition to the post-fossil age by a few decades. Ethicists agree almost unanimously that the protection of future generations requires the application of the precautionary principle.[5]

In this context, the hypothesis of a 'rich future', which states that the future will always be better than the present, is interesting. As early as 1785, Kant wrote the following on generational relationships: 'What remains disconcerting about all this is, firstly, that earlier generations seem to perform their laborious tasks only for the sake of later ones, so as to prepare for them a further stage from which they can raise still higher the structure intended by nature; and, secondly, that only later generations will in fact have the good fortune to inhabit the building on which a whole series of their forefathers (admittedly, without any conscious intention) had worked, without themselves having been able to share in the happiness they were preparing.'[6] Rawls also assumed an 'autonomous social savings rate'[7] and thereby a quasi-natural constant improvement of the living conditions of future generations. Caney explains what the 'rich future'-argument means with regard to climate change: 'The thought here is that future generations will be wealthier than current generations and hence more able to pay; as such an 'ability to pay' criterion should allocate duties to them. This, in effect, amounts to a policy of not preventing climate change for now and then trying at some point in the future both to prevent further climate change and also to adapt to the changes that have occurred.' [8] According to this view, the argument that because of the abrupt climate change future generations will be worse off than they would have been without it[9] has little weight. Because the lot of currently living generations *in sum* is worse than that of future generations, it would be unfair to demand a sacrifice from the current generation for the sake of future generations. The Danish economist Bjørn Lomborg writes: 'And by this time [2050] the average citizen of the world will have become twice as wealthy as she is now. (...) This total future consumption also underscores that global warming is not anywhere near the most important problem facing the world.'[10]

Now, what do we make out of this? There are indeed empirical facts that seem to support the theory that the future people will be better off than the present. For instance, the Human Development Index (HDI) has increased globally in recent decades despite the onset of climate change. For the *average* citizen of the world, who is the subject of intergenerational equity, per capita income, life expectancy and level of education are higher today than in the previous or pre-previous generations. The HDI of Bangladesh, particularly hard-hit by the greenhouse effect, rose between 1980 and 2011 even faster than average, from 0.303 to 0.500.

In spite of this, the 'rich future'-argument remains unconvincing, because it implicitly suggests that it would be fair if a future generation were exactly as well-off as its predecessor. It is crucial to understand that intergenerational justice means making possible not an equally good but rather a *better* life for future generations.[11] If our normative obligations to future generations are indeed greater than we commonly assume, the 'rich future'-argument loses its basis.

The concept of 'intergenerational justice as making improvement possible' implies that catastrophes that could lead to rapid and extensive losses for human well-being be pre-emptively averted by the current generation. It is therefore the most important duty of every generation to avoid war and environmental, social and technical catastrophes.[12] Climate change is one of the potential catastrophes that could descend on coming generations. The above cited assertions from Kant and Rawls concerning the betterment for posterity do not constitute laws of nature – on the contrary, the fate of coming generations hinges on our actions. In the case of climate change, future generations' costs for assimilation are presumably much higher than the current generation's costs for prevention. In fact, because of the tipping points depicted above, it is very probable that acclimation by financial means will not be possible at all, and the high numbers of dead and injured can only be deplored. As Caney points out,[13] it would be immoral to *knowingly* cause harm to future generations. The fact that someone is in a position to redeem himself, for example in the case of theft, to personally replace the stolen good, does not legitimise the theft itself. Discounting the future because it is far away or not here yet is inadmissible.[14] Temporal and spatial distances have a lot in common. Whether a murder occurs one mile or 100 miles away, whether it happens in one year or 100 years – it is still a murder and inherently wrong. In the same way, discounting the damages due to climate change in the distant future is morally inadmissible.[15]

8

PURE DISTRIBUTIVE JUSTICE

Keeping in mind our provisions for future generations, how can atmospheric resources be fairly distributed among members of present generations? According to the maxim of pure distributive justice (without further assumptions), the 'presumption for equality' applies: Everyone gets an equal share of the pie, i.e., every person has the same greenhouse gas emissions budget. For a (static) world population of approximately 7 billion people and a 'safe emissions budget' of 560 gigatons, this results in a per capita emissions budget of 2 t/year between 2010 and 2050.[1] As Figure 8.1 demonstrates, this concept calls for individuals with high consumption levels to adapt their behaviour to comply with an emissions limit of 2 t/year, while individuals with lower than 2 tons of emissions per year are allowed an increase. This is also the target value used in political negotiations to limit the average global temperature increase to 2°C.[2]

Dividing the limited budget by the number of individuals corresponds to the ethical stance of emissions/certificate egalitarianism. It grants each person on the planet an equal share of the atmospheric resources. This approach, also called 'Contraction & Convergence (C & C)', is currently the concept that is probably the most advocated in ethics.[3]

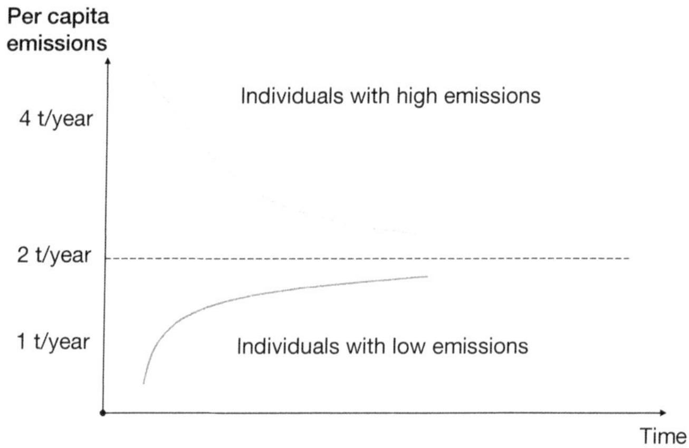

Figure 8.1 Contraction and conversion: emission egalitarianism assuming a constant population of 7 billion

According to Page, it 'seems congruent with both "contribution to problem" and "ability to pay" arguments for differential responsibility, yet it does not depend on either of these for its essential justification. The approach does not assume that those that must make the biggest changes in their environmental practices were responsible for the climate problem either historically or contemporarily.' Moreover, C & C 'seems consistent with a range of theories of the profile of justice. It will be attractive to egalitarians (...) as it will reduce inequalities between developing and developed countries, and between generations (...). It will also tend to improve, relative to rival approaches, the position of the worst off (...). Finally, it will be attractive to those who wish to bring as many people as possible to the point where they have enough since the measures it will introduce will benefit many millions of people in developed and developing countries who lead, or will lead, lives lacking in what is needed for a decent life without bringing more than a very limited number of people below the sufficiency level.'[4]

As a rejoinder, one could bring forth the 'climate region argument', contending that this allocation is too simplistic: people in temperate climates who, objectively speaking, need less energy for heating or cooling, are given the same budget as people who live at the equator or in polar regions. However, it would be incredibly complex to calculate how much more energy per person is needed for people in extreme climate zones than for those in moderate climate zones. After all, people often migrate during their lives. And moreover, climatic zones also change, precisely because of climate change!

A further objection is based on John Rawls' 'difference principle':[5] 'Social and economic inequalities (...) are to be to the greatest benefit of the least-advantaged members of society'.[6] For example, if everyone were compensated equally, who would expend the time and energy necessary for being educated as a doctor? In this case, differential salaries for such fields also benefit the society's weakest members. If we were to apply this to 'climate logic', this would mean that resource- or emissions-intensive professions may sometimes benefit all of humanity. In these cases, according to the difference principle, it would be allowed, indeed morally imperative, to deviate from the principle of egalitarianism. Like the 'climate region argument', this argument based on the difference principle is intellectually appealing but hardly viable on a practical level. The actors requiring more emissions than average would have the burden of proof. An audit authority would have to be established to determine whether their CO_2e-intensive activities actually benefit the weakest members of society. That seems hardly feasible.

Prioritarianism, on the other hand, awards a higher CO_2e budget to people with little wealth or resources *precisely because they are poor.* Several NGOs as well as developing countries put forth the 'Greenhouse Development Rights

Model'[7] as an alternative to 'Contraction and Convergence', because it combines development aid with climate-related adaption aid for the Less Developed Countries. They say that 'Contraction and Convergence' alone is not fair enough, and furthermore, with their model one can kill two birds with one stone.[8] This, however, confounds the climate problem and the general problem of poverty, thus confounding two different reasons for redistribution between the rich global North and the poor global South. We focus here on climate-ethical considerations. Poverty, sickness, malnourishment and high infant-mortality rates are important concerns for the international justice debate and were so before climate change became an issue. We see no necessity to confound these issues with climate ethics theories. A fair climate change solution would not be a panacea for all problems of international justice.[9]

Given these inconsistencies of prioritarianism, the concept of 'Contraction & Convergence' seems to be the better vantage point for considerations concerning the distribution of greenhouse gas emissions. Even C&C is too ambitious for politicians of some rich countries. One need only recall the now famous words of the then US President Bush at the world summit for environment and development in Rio in 1992 when he said the 'American way of life' is not up for negotiation.

Politically, the question of the transition timetable is highly controversial. From an ethical point of view it should be relatively short, because circumstances deemed fundamentally immoral should be remedied as quickly as possible. The argument that some individuals who currently consume at rates much higher than the 2 tons/year goal would have to rapidly and radically alter their lifestyle is a correct but for ethics inconsequential counterargument.

9

INTERNATIONAL JUSTICE

International justice addresses justice between different countries, regardless of distribution practices within the respective countries. Nations and their political leaders, not individual citizens, make decisions about emission rights in the international arena. Pure distributive justice (see above) could only be applied if we had one world government that had the power to allocate resources fairly among its citizens. In reality, however, the international domain is divided into countries vying for influence and in possession of various bargaining powers, many of whom strictly pursue national interests.

As Figure 9.1 shows, according to emission egalitarianism within the concept of international justice, high-emission *countries* and low-emission *countries* should meet in the middle. Applying the concept at a national level does not initially change the objective (2 t/capita/year). States whose citizens currently release too many greenhouse gases into the atmosphere must reduce their emissions. States with low emissions levels have room to breathe – or emit.

As a rejoinder, one could argue in favour of a 'Rawlsian difference principle' *on a country level*. According to this view, we should first ask what the high-energy usage in developed countries is accomplishing – in particular for underdeveloped

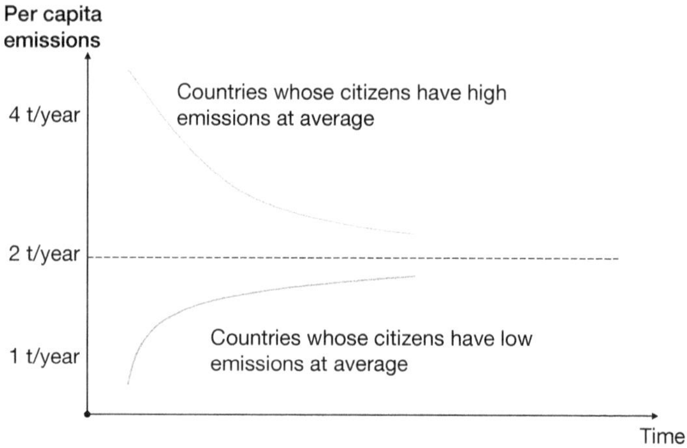

Figure 9.1 Distributional justice between countries (international justice) according to the principle of equal distribution at a constant population

countries. One could argue that the high energy use in the developed countries supports scientific and technological R&D which, with wise policies, might provide solutions to the energy and population based emergencies ahead. It would follow that there may be an 'international difference principle' at work, whereby unequal national per capita energy use is justifiable if that unequal distribution works to the advantage of the low-use nations. As mentioned in the section on pure distributive justice, we find this thought plausible, but we believe that the burden of proof lies with the high-emitting country. At the moment, these countries do not appear to be using their extra emissions to the benefit of mankind, rather they maximise their own well-being at the expense of other nations.[1]

Figure 9.2 shows carbon dioxide emissions of various countries per person per year. Of the industrialised countries, the United States leads with 19.74 t/capita, followed by Australia (19 t/person) and Canada (17.91 t/person). Germany is right around 10 t/person, while Spain and Italy emit

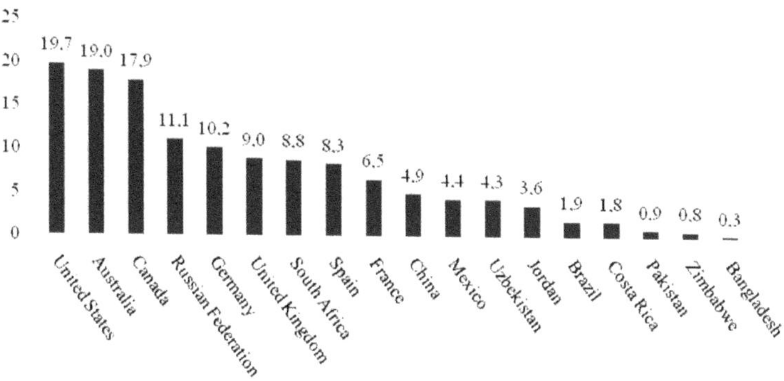

Figure 9.2 Carbon dioxide emissions per capita per year compared by country (2007)[2]

approximately 8 t/capita, France 6.5 and China just under 5 t/capita. It is important for the political debate that many developing countries also exceed the 2 t/person allowed according to C&C. The average for people living in Mexico is 4.4 t/capita; in Thailand average emissions are 4.1 t/person, and even in Jordan emission levels average 3.6 t/capita. Those under the 2 t/capita limit are the citizens of approximately 100, predominantly African, South American or Southeast Asian countries, e.g., the people of Costa Rica (1.8 t/person), Zimbabwe (0.8 t/person) or Bangladesh (0.3 t/person).

If we take a look at the *total* emissions (instead of per capita emissions) the picture is quite different as Figure 9.3 shows. China has overtaken the US a few years ago and is now the greatest emitter of the world with 24 per cent of world emissions. The US comes second with 18 per cent of global emissions, followed by the whole EU (27 countries) with 14 per cent, India with 6 per cent, Russia with 5 per cent and Japan with 4 per cent. The states that have accepted a climate regime (Kyoto II) account for only 15 per cent of global emissions and their share of world emissions is constantly shrinking.

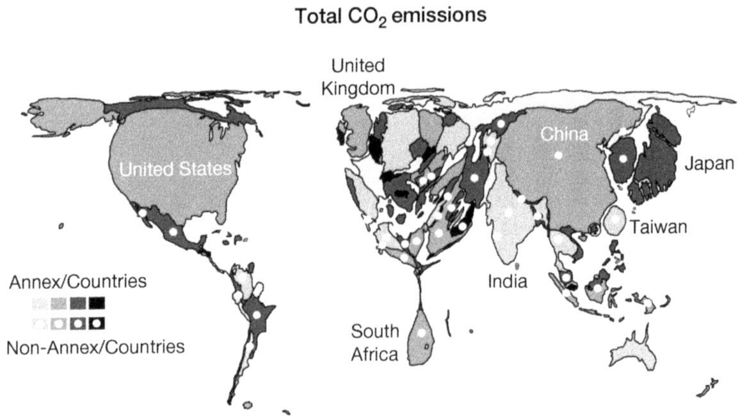

Figure 9.3 Carbon dioxide emissions compared by country (2004)[3]

It is important to understand that implementing international justice can conflict with principles of pure distributive justice. Let us assume that a citizen of a high emissions country lives extremely energy consciously. She foregoes trips that require flying,[4] commutes only by bike and seldom turns the heater on in the winter. Because this person has the misfortune to live in a country where the majority of her fellow citizens waste a lot of energy, then she might, in the context of a climate treaty based solely on the concept of international justice, nonetheless have to participate in a financial transfer from her country to developing countries, at least in some institutional designs. This contradicts the principle of pure distributive justice.

International justice in a world with population growth

The amount of CO_2 emissions is not only about the output of each individual, but also how many people live on earth altogether. The Fourth Assessment Report of IPCC

(2007) states: '[G]ross domestic product per capita and population growth were the main drivers of the increase in global emissions during the last three decades of the 20th century.'[5]

In the twentieth century, the world population rose from 1.6 to 6.1 billion, i.e., it almost quadrupled. The biggest part of this increase – 80 per cent – took place in the second half of the twentieth century. Now, there are more than 7 billion people on earth.[6]

According to the UNFPA, 40 to 60 per cent of the rise in CO_2e emissions can be attributed to population growth.[7] And the world population is predicted to grow from 7 billion now to 9 billion in 2050. In terms of absolute emissions, this means that countries with a growing population will generate an ever-growing proportion of global emissions, whereas the share of total emissions of nations with decreasing population trends will decrease. Table 1 illustrates this phenomenon by comparing population trends in Germany and Pakistan. By 2050, absolute emissions of each country are almost equal.

Can or should population size be taken into account? With regard to intergenerational justice, the answer is yes.

Table 9.1 Development of the absolute climate gas emissions of Germany and Pakistan

Germany	2010	2020	2030	2040	2050
CO_2/person/year	9.5	9.5	9.5	9.5	9.5
Population (in millions)	82	80	78	74	71
Absolute CO_2	779	760	741	703	675
Pakistan	**2010**	**2020**	**2030**	**2040**	**2050**
CO_2/person/year	1.8	1.8	1.8	1.8	1.8
Population (in millions)	185	226	265	302	335
Absolute CO_2	333	407	477	544	603

Source: UN Population Division (2009).[8]

It is clear that given current population growth trends, the environmentally sustainable amount of greenhouse gas emissions is below the current limit of 2 tons per person.

In addition to the consideration of population growth in the world as a whole, the ethical-political question arises as to whether population growth in *individual* countries should be taken into account in determining emission limits. It is conceivable to 'reward' a country with a constant or declining population by allowing it to produce more greenhouse gas emissions than a country with rapid population growth. The acceptability of factoring in population changes largely depends on our general stance on population policy. Political theorists that view population policy as something that is not and should not be influenced by politics would argue against including population as a variable in determining emission limits of individual nations. If population policy, however, is seen as ethically and politically admissible, then it is appropriate to factor in population trends in the determination of national emission budgets. The population policy debate is too complex to be included here, but as argued elsewhere, not all population policies are immoral *per se*. On the contrary, the state is free, within reason, to implement incentives to promote an anti-natalistic approach to fertility among its citizens.[9]

In Figure 9.4 it is assumed that population is an exogenous variable that is not and should not be influenced by governments. Those countries which presently consume a lot must therefore reduce their emissions even more drastically than in Figure 9.1 which assumes a static population. The leeway for increased emissions in countries currently consuming less would also be diminished. With the premise that the safe emissions budget remains 560 Gt, the morally legitimate per capital emissions move towards 1.5 t/capita.[10]

In Figure 9.5 it is assumed that a government can and should influence its population size, which in turn becomes

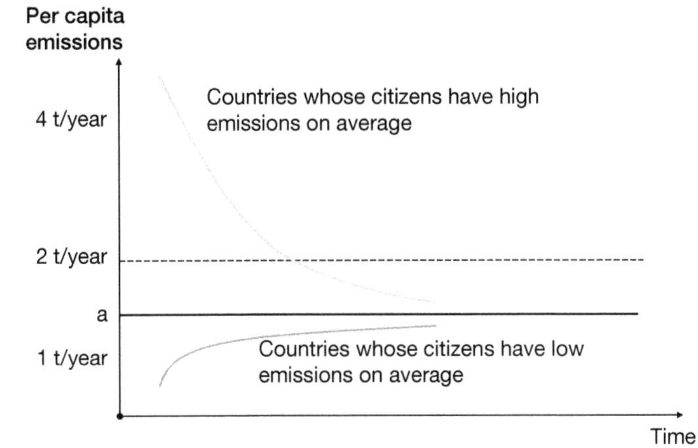

Figure 9.4 International justice given a population of 9 billion people and an ethical inadmissibility of population policy

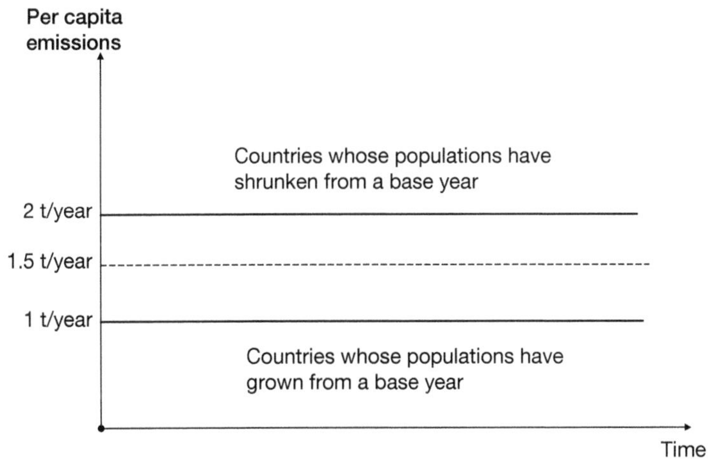

Figure 9.5 International justice at a population of 9 billion people and an ethical admissibility of population policy

an endogenous variable. Should the allocation of emission allowances occur today (2011), with a world population of 7 billion people, countries with subsequently constant population counts would have in a world with 9 billion people in 2050 more emission allowances per capita than those of growing countries. If the average morally legitimate emission budget would be, say 1.5 t/person in 2050, countries whose populations have grown from now would have for instance 1 t/person whereas countries whose populations have shrunken would be allocated 2 t/person. If the distribution were to take place at an earlier date, 1990 for example, the ratio would shift further to the advantage of those countries whose population has not grown or even shrunk since then. In any case, population changes going forward from some base year are relevant to how emission budgets should be distributed.

10

HISTORICAL JUSTICE

Historical justice is often grouped with intergenerational justice since both concepts deal with 'justice in time'. This, however, is an analytical deficiency. Historical justice divides a generation into groups, namely, at least one injured party and one at-fault party. Intergenerational justice, on the other hand, deals with justice between members of different generations each represented by an *average* individual of its epoch.

The ability to produce the natural resources coal, oil and gas on a large scale made the Industrial Revolution possible; however, this did not take place everywhere at the same time. Rather, industrialisation arose in various (groups of) countries at different points in history. As Figure 10.1 shows, the cumulative historical emissions of greenhouse gases, i.e., those generated since 1850 are much higher in North America, Europe and the former Soviet bloc than in Africa or Latin America.

Between 1850 and 2002, industrialised countries emitted three times as many green house gases as developing countries.[1] The amount of greenhouse gases in the atmosphere have increased from 280 ppm to 386 ppm since the beginning of the Industrial Revolution. Consequently, more than half of the sanctioned greenhouse gas budget for all

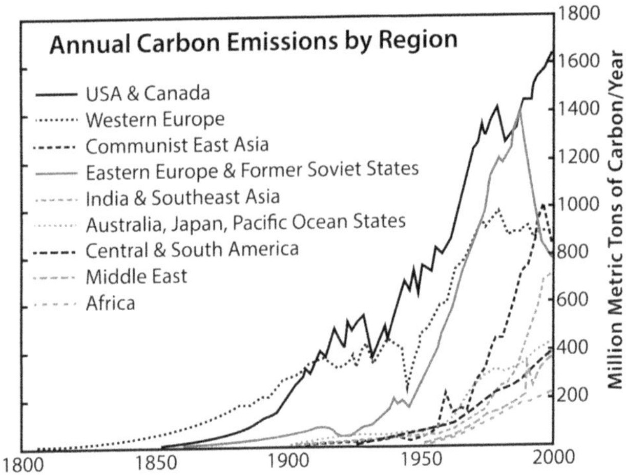

Figure 10.1 Historical emissions by region

mankind according to the 560 Gt CO_2e or 2 degrees Celsius target has already been consumed, and industrialised countries (the North) are responsible for the lion's share of this consumption. Based on this, one could support a right to 'overshoot' for all-time low emission countries.[2] As shown in Figure 10.2, the citizens of all-time low emission countries would be allowed to exceed the level of 2 t/year initially, but then must converge down to this mark. Countries with the most all-time emissions, on the other hand, would have to significantly reduce their consumption below the level of 2 t/year. They may raise their emissions later – if the historical encroachments of their countries have been compensated for over time – to the level of 2 t/year.

Historical justice is (thus far) a highly controversial ethical concept in which there is much less consensus among ethicists than in many other areas. 'Historical *in*justice' is usually defined as a violation of the rights/interests of deceased people by other departed. Neither those currently claiming injustice nor those accused were involved, but rather

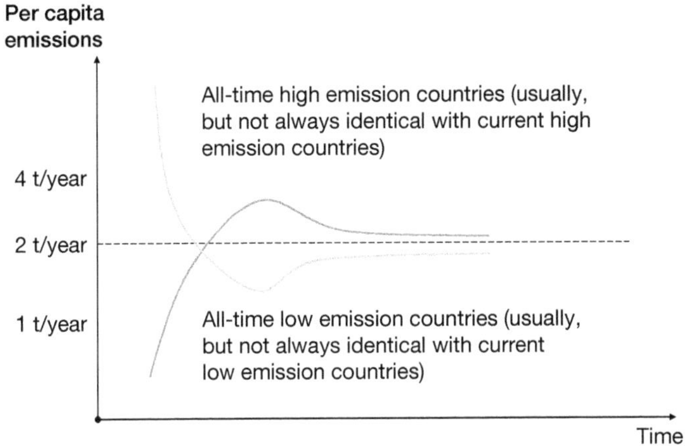

Figure 10.2 Historical justice, 'overshooting' of the countries with low accumulated emissions

are descendants of the perpetrators or victims.[3] Key questions of 'historical (in)justice' are: How should we deal with immoral deeds of deceased individuals, the negative effects of which are felt to this day, especially such acts committed under the aegis of a rogue regime?[4] Do the descendants of the by-this-time-deceased perpetrators still have obligations to the descendants of the victims of historical misdeeds, for example the return of stolen objects (usually territories) or compensatory payments? And how far back in history can we go to atone for past misdeeds?

The question arises in general as to what extent the concept of historical justice can reasonably be applied to climate change. Thought experiments are commonly used as a method of analysis in ethics as they use well-established concepts, the logical deductions of which we know and trust, and apply them to newer, more complex issues. Let us discuss two analogies that facilitate the developing of ethical conclusions.

Analogy 1: Two hikers have a ration of ten litres of water and 20 apples available for a walk. Halfway through the hike, one traveller (the North) has consumed half of the rations. Now he begins aggressive negotiations with the other traveller (the South) to decide how the remaining five litres of water and ten apples should be allocated. It is understandable that dividing the rest of the rations *equally* seems completely unfair to the hiker who has not yet had anything to eat or drink.

Consider, though, a reverse analogy:

Analogy 2: A hiker is alone on a trip and gets hungry. When he discovers an apple tree, he shakes the ripe apples from the tree and eats until he is full. A little while later, a second hiker arrives at the same place and complains that he cannot stave off his hunger. He would have liked to have had half of the apples. In this example, no one would claim that the second hiker had a right to the apples from the start and can therefore demand compensation.

Imagine two generations of Americans and Pakistanis – at one time the two peoples were living at the end of the nineteenth century during the Industrial Revolution, at the other time in the present.[5] Let us suppose that the Pakistani generation in the nineteenth century produced virtually no greenhouse gases; the American generation, however, did. Let us further assume that the Americans derived benefits from their CO_2e production, e.g., in the form of a significant increase in the standard of living. This advantage was passed down through generations, so that it also guarantees Americans today a level of prosperity well above that of the Pakistanis. Let us further assume that the historical emissions of the deceased Americans are directly responsible for damages concerning today's living Pakistanis, for instance through flooding or crop failures. Do the current US citizens owe the current population of Pakistan compensation for the emissions of their American ancestors? Should the present inhabitants of the largest polluting nations compensate the victims of the greenhouse gas emissions of their ancestors generally?

Before we begin to answer these questions, a few of Gosseries' assumptions should be formalised and clarified. The relationship of well-being between two parties can exist in the following forms:

1) Actions that increase the well-being of party A while diminishing the well-being of party B (win/lose).
2) Actions that increase the well-being of party A and have no effect on the welfare of party B (win/0).
3) Actions that have no effect on the well-being of party A but diminish that of party B (0/lose).

The ethically relevant cases are those that actually caused damage to another party (therefore scenarios 1 and 3). Actions that increase the well-being of party A and have no effect on the welfare of party B (scenario 2) do not belong to the realm of compensatory justice.

First of all, we must clarify whether a lose-situation exists in conjunction with climate change and if so, to whom and to what extent. According to current research in the natural sciences, historical emissions are at least partially causally responsible for current natural disasters – this is an example of number 3 (0/lose), as affected regions undoubtedly incur damages due to such incidents.

From a moral point of view, however, these cases of loss are fundamentally different from the distribution of the absorptive capacity of the atmosphere. This will be made clear by a further apple tree analogy:

Apple tree analogy 3: As in our second thought experiment, a solitary hiker happens upon an apple tree situated on a rock. He would like to satiate his hunger and begins to harvest the apples. In the process, he shakes the tree and unintentionally sends an apple rolling. With great momentum, it falls on the head of a second hiker who was much further down than Hiker 1. This injures the second hiker quite badly.

The first hiker now has not just one but two moral issues to resolve with the second hiker. First, he (without guilt) ensured that the second hiker had nothing to eat (a question of distributional justice). Second, he (likewise without guilt) inflicted the pain on the other (a question of compensatory justice). In the case of climate change, these two issues must be handled separately as well. It is not helpful to couple both concepts in statements like, 'The North filled up the atmosphere with its greenhouse gases more than was admissible, and now the South has to endure the consequences in that it especially suffers under global warming and heavy flooding'.

What, then, is the status of the distribution of the 'absorptive capacity' of the atmosphere? Is the historical depletion since 1850 an example of case 2 (win/0), that is, that A's situation improves while B incurs no damages? Or does the use of fossil fuels by nations industrialised early on illustrate case 1 (win/lose)? Analogy 2 (see above) is helpful here: Someone who discovers and uses a resource on his own, in my example an apple tree, does not prima facie inflict harm on anyone else. It is an objective fact, however, that the second hiker B feels hungry and cannot eat precisely because hiker A satiated his hunger a few hours earlier. A has worsened B's situation, even if B does not blame A. The worsening has happened even if B assures that he would have done the same in A's position. As it concerns the absorptive capacity of the atmosphere, even if earlier generations' consumption thereof was legitimate at the time of use, the current situation does not fulfil John Locke's dictum, 'as much and of the same quality may remain.'[6] The matter is not one of a direct, but rather an *indirect* worsening of the situation for those who came too late. In conclusion, we are faced with a win/lose situation and therefore one that falls into the jurisdiction of the ethicists.

In the following, we will solely consider whether the global North is obligated to offer some sort of compensation for its early access to the 'absorptive capacity' of the atmosphere. What counterarguments may be cited?[7]

Firstly, the general question arises as to what extent responsibility or fault can be carried over from an earlier to a later generation. Let's assume I find some bottles of wine in a box in the basement of my newly acquired house. If I want to distribute them among my five friends, and if I find out that the now deceased grandfather of one friend had already taken one of the original stock and consumed it, do I then give his grandson one bottle less? Couched in these terms, it sounds unfair. However: Assume A1 steals a bar of gold from B1 and wills the money to his only son A2. Upon discovery, B2, the son of B1, could then demand the return of the bar of gold or compensation, even if A1 and B1 have already died. This applies regardless of whether A2 knew or should have known what his father did. The only decisive factor is that he benefits at present from the money he inherited.

The advantage of the second example, of course, is that the contended object can be returned, an option ruled out in the first example (and in the case of emitted greenhouse gases). Additionally, the damages incurred by the person from whom was stolen are exactly quantifiable (in marked contrast to damages due to climate change).

Material (though guiltless) responsibility has to be distinguished from a moral responsibility. But most liberals deny the inheritance of *any* responsibility, or desert, on the individual level, from father to son, from mother to daughter.

In liberal concepts, the inheritance of blame is considered illegitimate not only on the individual basis but also, and even more, on a country level. Each generation of citizens is to be regarded as his/her own 'demos' and not as part of a generational chain. On the other hand, some other theorists (especially communitarians) do believe that citizens as members of enduring states are obliged to compensate misdeeds that were committed in the past in the name of the same state. According to this view, descendants of slaves have legitimate claims against the enduring legal successors of former slave states, provided amends have not been made for misdeeds committed against their forefathers.[8] But, as mentioned, this is highly controversial.[9]

Another argument against any kind of compensation for historical emissions is the *innocuousness* of the burning of fossil fuels, i.e., the absence of immorality in the act of utilisation. The example of slavery differs from that of greenhouse gas emissions first and foremost because the slaveholders transgressed basic moral obligations towards their fellow human beings at the time of the deeds. In 1850, burning fossil fuels was a win/0-action. In contrast, slavery was a win/lose-action in 1850. Even today, the burning of fossil fuels is an action that is not deemed *inherently* bad.[10] A warming fire, a car drive, even a flight are all decidedly different from murder, theft and homicide. Only for the latter group of undertakings does the term 'reparations' seem appropriate. For this reason, some scholars advocate a trans-generational obligation to reparations for slavery and the holocaust but rejects such an obligation in conjunction with the burning of fossil fuels.[11]

A further aspect of innocuousness, largely ignored in the literature, is the evaporation of greenhouse gases over time. When calculating historical emissions, it makes a difference whether every ton is equally weighted or if one considers that some greenhouse gases have already partially evaporated, thus making the countries' share of the current concentration of greenhouse gases in the atmosphere different than their share of historical emissions. According to IPCC, 50 per cent of CO_2 evaporates within 100 years. Methane stays in the atmosphere around eight years before it is evaporated.[12]

A related but not identical counterargument against compensations, thus the third, is the *ignorance* of the consumers. At the time of the Industrial Revolution, climate science did not exist, and the global warming effect of carbon dioxide and other gases was unknown. By degrees, scientists investigated the essential correlations of the anthropogenic climate change. In the first IPCC report in 1990, the majority of scientists around the world supported the hypothesis that the current climate change was man-made. This hypothesis was solidified in the second IPCC report in 1995. Today, the accumulation of CO_2e in the atmosphere is no longer seriously doubted.[13] Therefore, the publication of the first IPCC report in 1990 can be seen as the point in time from which the argument of ignorance is no longer valid.[14]

Changing national borders is also a strong counterargument to the concept of the inheritance of guilt or responsibility. Let's assume that the Soviet Union emitted an exceptionally large amount of greenhouse gases in the region that is now Uzbekistan. Should these emissions today be apportioned to the Uzbek state, which was not founded until 1991? This seems unfair. A further problem emerges when a land was not democratically governed at the time of its emissions: South Africa is admittedly a substantial all-time emitter. However, the apartheid government is responsible for these politics – the black population profited little from the use of fossil fuels. Should the now democratic government be held responsible for all of South Africa's historical emissions? Hardly. These examples show that the current demos must not take the responsibility for the climate gas emissions of earlier generations of inhabitants of one's own present country in every case.

A final argument: The premise for any legitimate claims of the South against the North is a betterment or advantage for the currently living due to the historical climate gas emissions of the North. If current citizens in industrialised nations have not profited from the undertakings of their ancestors, then people in developing countries have no rights to reparations. The essential question of whether people living today have an advantage as a result of historical emissions is ultimately an empirical question: Can the developmental head start of certain countries be attributed solely (or at least, mainly) to their copious greenhouse gas emissions in the past? Or do other factors such as work ethics and diligence, dexterity and creativity play a more important role? Can the comparatively slow development of other countries be traced back to low past levels of greenhouse gas emissions? Or are other factors (partially) responsible? Several ethicists assume a causal correlation between quality of life and historical emissions without investigating their assumptions:

'And since the wealth of the developed nations is inextricably tied to their prodigious use of carbon fuels (a use that began over 200 years ago and continues unchecked today), it is a small step from here to the conclusion that the present global distribution of wealth is the result of the wrongful expropriation by a small fraction of the world's population of a resource that belongs to all human beings in common.'[15]

'In this sense, climate change is a by-product of the affluence of the world's most advantaged nations and persons.'[16]

We do not deny a tangential correlation between the speed of development and the use of fossil fuels. In pre-industrial societies, energy was a good that was only available in acute scarcity. Even at the end of the eighteenth century, 40 per cent of the total usable energy came from the muscular power of draft animals. Another 40 per cent came from logging, 12 per cent from water wheels, 3 per cent from human labour and less than 1 per cent from wind mills and coal.[17] In Europe, this constellation changed radically with the onset of the Industrial Revolution. Thanks to the versatile uses of coal, especially in the steam engines powered by coal in factories, a metal and engineering industry could now be developed and numerous products produced industrially. The railways and steam ships, which could also now be built, transported these goods all over the world. 'Thus, the use of fossil fuels in combustion engines became the motor of globalization, which first found an energy-based economy in the colonial era of the 19th Century', writes the climate-historian Ludwig.[18] However, Jared Diamond (1997) tells a different story about the sources of the wealth of nations, underlining the role of germs and diseases.

Singer's and Vanderheiden's argument is based on the questionable assumption that technological developments have *only* benefitted the North. But in a world in which goods can easily transferred from one country to another, a solar cell or a pharmaceutical drug can also benefit other people than the inventors.

To sum this fourth counterargument up, a lot of empirical research still needs to be done, but it seems as if there is no simple linear relationship between a country's historical emissions and its citizens' current quality of life. If however, empirical research were to show that this correlation is strong, the following analogy would apply:

Hiker A and Hiker B are hungry and go for a walk. They see an apple tree. Hiker A has a basket and fills it up with apples, Hiker B has no basket and is only able to take enough apples to cover the energy used for the walk. Once back in the village Hiker A and his daughter eat half the apples and Hiker A saves the other half for later. Hiker B takes the hint and spends the evening making a basket. Unfortunately, both Hiker A and Hiker B die during the night and leave all they have to their daughters. Daughter A gets as basket and half a basketful of apples while Daughter B gets a basket. Daughter A and Daughter B go on a walk with their baskets. They see a second apple tree. They come to learn that there are only two apple trees in existence (making this second tree also the last one).

In this case, it would be fair to let Daughter B alone harvest the second apple tree.

What is the upshot regarding historical justice? The weight of the four counterarguments (only the fourth yielding an unclear result) against claims based on historical emissions is so considerable that there is no moral *obligation* for the North to compensate the South for initial access to the scarce resource 'atmosphere' (before 1990). The concept of retributive compensation does not apply. Countries with low historical emissions have no rights to reparations, since they were not wronged in any way. They were harmed, but not wronged.[19]

Caney (2010b) or Meyer/Roser (2010) arrive more or less at the same conclusion. Meyer/Roser[20] restrict it to 'those emissions that belonged to people who are now dead and which yield no benefits for the currently living'. Regarding the emissions that do yield benefits for the currently living, Meyer and Roser hold that these benefits have to be shared between the North and the South because justice (distributive justice, not compensatory justice, but still *justice*) demands this from us. We advocate the view that the idea of justice is not suitable for this cause, but that supererogatory duties like solidarity or charity or benevolence demand us to compensate the South for the North's excessive use of atmospheric resources before 1990. Regarding the term of 'justice', a few points should be noted:

Morality is not exhausted merely in complying with mandates of justice. The scope of morality also encompasses good-naturedness, benevolence, sympathy, compassion, altruism, generosity, mercy, solidarity, empathy, charity and other such qualities. Justice is only a part of morality, albeit a very important one. Whoever is not empathetic, generous, and charitable will disappoint others, but whoever is unjust will outrage them. Public opinion can only demand what we owe each other, and that is justice. Meritorious conduct that goes beyond that is up to each individual.[21] There is no moral obligation to compassion, whereas it would be immoral not to fulfil obligations of justice.

Secondly, justice is sometimes referred to as the 'totality of enforceable standards', whereas the rest of morality is called 'the totality of recommended standards'.[22] But that is not correct as such: enforceability is not a constitutive property of justice. For instance, the racial segregation in the USA from 1865 until the Civil Rights Act in the 1960s was certainly unjust, but the laws of those days allowed it, so its abolishment was not enforceable.

Thirdly, there are circumstances of justice. One precondition is that humans – in relation to each other – desire scarce goods.[23] Robinson Crusoe cannot have a justice problem, as the smallest scale on which such a problem can arise is with at least two people.[24] Two men want to marry the same woman, two generations want to use the same fossil resource. Principles of justice are only necessary where people have diverging – and potentially conflicting – interests, and goods are not abundant enough to fulfil them all.

Fourthly, the question of a just social order is a problem of secular-liberal societies. Religious societies have a pre-set order. Everyone has his predefined place in the world. Whatever has been decided by God, or the Gods, is just. In the Middle Ages, Europe was characterised by such a static order. Only when this order was abolished during the Enlightenment, justice had to be determined by man himself. All of a sudden, the world was seen as the product of human action. Hence, moral principles (including those of justice) had to be defined and established by man.

All this shows that a ubiquitous use of the term of 'justice' should be avoided. Let us come back to analogy 2. It was assumed that hiker B did not find any more apples because hiker A had consumed them earlier (without incurring any guilt). Hiker B is at the same point as he would have been had the apples not been available due to a natural cause, e.g., they rotted before he got there or lightning destroyed the apple tree before the hiker discovered it. Tough luck, one could say. But not a question of justice.

This, however, is not the end of the story. Analogy 2 can be told further: After the initial event (A filled himself up on apples, B did not), the two hikers decide to go forward together. In the evening, both are hungry again (B more so than A). They find a second apple tree. A could now let B eat more than half of the apples. This would be a generous, benevolent gesture though not imperative for the sake of fairness. One hears often from developing countries that countries industrialised early on should cut back on future utilisation of emissions so the developing countries can 'catch up'. Hiker B's higher consumption in the apple tree example corresponds to the overshooting that developing countries would like. There is much to be said for at least partially fulfilling their request out of altruism.

Even if one rejects the notion of greenhouse gas emissions as a case of historical injustice, the question of current emissions remains. Strictly speaking, this is no longer a question of historical justice, because the causal agents as well as those suffering belong to the currently living generation. For emissions after 1990, pleading ignorance is no longer a viable argument. As the emitters are still alive, the controversial discussion about the inheritance of responsibility does not apply. For emissions after 1990, Meyer proposes that those who have over consumed up to now have fewer emission rights in their remaining lifetime.[25] Those who consumed a lot early have few remaining emissions and vice versa. Conveniently, in the case of greenhouse gases the solution identified as 'fair' can be based on a person's entire lifespan.[26] From an ethical perspective, therefore, there is a good case for an 'undershooting' of the citizens of industrialised countries for the rest of their lives.

11

FROM THE CURRENCY OF GREENHOUSE GASES TO QUESTIONS OF MONETARY DISTRIBUTION

Four concepts of justice (intergenerational, pure distributive, international, historical) have now been applied to the climate problem. Perhaps it has occurred to the reader that our ethical conclusions up to this point have not addressed the *amount* of financial compensation that the North owes the South. Indeed, we find the tendency of many climate ethics essays to go from the currency of greenhouse gases to questions of monetary distribution overly hasty, because the fair distribution of a scarce physical resource is not initially a question of financial payments. Rather, it is a question of the distribution of the scarce resource itself.

Nevertheless, to what extent does a 'fair' solution to the climate problems mean financial transfers between (groups of) people and in what amount?

In the language of the Intergovernmental Panel on Climate Change (IPCC), 'mitigation' and 'adaptation' are differentiated. Even without getting into complicated opportunity cost calculations,[1] it can be said that both activities occasion costs. The costs of 'adaptation' accrue, for example, in the construction of dams, the relocation of people in coastal areas, irrigation systems for regions with less precipitation than before, modification of cultivation, the rescue of animal species whose existence is threatened because of climate change and the like. Costs for 'mitigation' arise, for example, from the retrofitting of factories with carbon filters or better insulation in buildings. However, greenhouse gases are above all avoidable through dispensation, for example dispensing with flying, driving, consuming meat, and finally having children – these forms of dispensation can hardly be described as 'costs'. But admittedly, mitigation accrues *some* financial losses, even if they are probably lower than the costs for adaption.

In the ethics (and political) debates, it is not the payments arising in agreed climate trading schemes, but the additional cost *transfers* that are controversial. For this reason, the greenhouse gas certificate trade will be excluded hereafter. For reasons of economic efficiency, greenhouse gases should be prevented on a global scale where the costs are the lowest. To this end, a developing country can sell its emissions' certificates to industrialised countries. This is the underlying idea of the certificates trade. And it is fundamentally correct, at least providing that the initial distribution (grandfathering) was agreed upon to the satisfaction of all, the market works, and abuse of the system (free-riding, outright fraught etc.) is prevented. Providing all this, how existing rights of developing countries are to be honoured – whether through actual additional emissions or through the sale of emissions rights – is an economic, not a moral question. Invoices for purchased certificates have to been paid, just like invoices for milk or butter. Ethics can confine itself to resolving the more complicated matters. In this context, one can fundamentally differentiate between:

a) Payment claims that are historically founded.
b) Payment claims that are founded with an eye towards the future.

Historically founded financial claims

Historically founded financial claims can be made for the use of the 'absorptive capacity' of the atmosphere. Depending on when this use took place, they should be treated differently.

1) Before 1990: The concept of historical justice is grounded in the opinion that the North has fewer rights to the remaining absorption capacity of the atmosphere because of its historical emissions. If one takes this position, two conclusions can be inferred that are frequently compounded in the political debate but that, analytically, should be clearly differentiated. The first conclusion, developing countries' right to overshoot, has already been thoroughly explicated. The second conclusion, the demand for financial compensation for the 'all time' low-emitting countries is an argument that must be differentiated from the overshooting. In any case, the developing countries have no rights in this respect, the industrialised countries have no moral obligations. Industrialised countries can voluntarily provide such payments out of compassion, but they do not owe them to developing countries on the grounds of justice.

2) Between 1990 and the present: For this time span, a moral right of the developing countries to compensation was justified above. The North owe it to developing countries on the grounds of justice. Whether this should be rendered through overshooting *or* money will be discussed later. To demand *both* is not fair.

3) For already eventuated direct damages: First of all, one must stress that the case of the fair distribution of the 'absorptive capacity' of the atmosphere (case 1 of the win/lose scenarios presented above) must be analytically differentiated from case 3 of the win/lose scenarios presented above, that is the *direct* damages, for example, due to extreme weather events brought about by climate change. If one Country A impairs another Country B through an unintentionally triggered meteorological event, we are dealing with an analogy to an unintentional bomb strike that, for the sake of the argument, has the same number of victims.[2] The overall considerations concerning the fair use of resources are a different subject, because they refer to *indirect* damages.[3] In the case of *direct* damages, the 'polluter pays principle' – a backward-looking principle – applies.[4] The fair distribution of the 'absorptive capacity' of the atmosphere can be done without changing the physical currency, that is: it was theoretically not necessary to introduce 'money' into a fair distribution scheme. In contrast, a direct damage to another country and/or its citizens induced for example by extreme weather events cannot be reversed by a redistribution of the 'absorptive capacity' of the atmosphere or any other physical unit. But can money reverse such damage? The application of the polluter pays principle in the context of climate change poses practical problems. The research in the natural sciences typically delineates a spectrum in which, between 1990 and 2100, the sea level will rise between 0.75 metres and 1.9 metres.[5] One can obviously identify this as a causal factor for current and future flood disasters. For a legal assessment of the damages, however, this is inadequate. The attribution to raising sea levels to damages is economically possible in principle. In order to establish an exact amount in Dollars, Euros or Pounds to which those who incurred damage are entitled in court, however, one must also investigate

the share of individual greenhouse gas emitters in particular damaging events. Admittedly, this seems unimaginable at the current state of the economics of climate science. This does not, however, weigh against the polluter pays principle itself. Instead of not using it at all, it should be applied with as much exactness as is currently possible. Caney argues that many polluters are already deceased and it would be unfair to transfer the costs of the effects of their emissions to their descendants.[6] This supports the argument that only damages from hurricanes, droughts, etc. resulting from the emissions of people who are still alive should be compensated through financial transfers from the polluter to the victim.

Different from the 'ability to pay principle' discussed below, the 'polluter pays principle' does not take into account how rich or developed the emitter is. China, as the country with the highest absolute greenhouse gas emissions in the world, would have to pay a significant portion of the global damages of climate change if the 'polluter pays principle' alone were to be used.

Financial claims that are founded with an eye towards the future

1) According to the principle of intergenerational justice, all greenhouse gas emitters should provide today for generations of the future in their own respective countries. This initially applies without a redistribution process between countries.

2) For geographic and economic reasons, future generations in parts of Africa, Asia and South America are more severely affected by the consequences of the abrupt climate change than future generations in industrialised countries. In addition to their own costs for measures of conservation and adaptation, industrialised countries, out of solidarity

or compassion (not for reasons of justice), should help the people/countries in these regions finance their measures of conservation and adaptation. Imagine that in a parallel world, an abrupt and dangerous climate change takes place in which *humans played no role*. The poorest suffer especially. Their huts are destroyed; the number of infections transmitted over water increases, their situation with regards to nutrition deteriorates. In this situation, it is a moral imperative to help. As with all similarly allotted payments for famine and natural disasters, the rich countries (and in them, in turn, the richer individuals) have the highest obligation. Caney speaks of the 'ability to pay principle' that would apply here.[7] In contrast to the 'polluter pays principle', it does not take into account at all who caused the damage, rather, solely who is able to pay. The corresponding to the 'ability to pay principle' is the 'principle of need', which is regulated by the receiving side: the needier a potential recipient is, the more he/she will be considered in the payments.

What about the use of moral arguments by political actors? In the next section, the extent to which the four ethical concepts presented here are used de facto in the political debate will be examined. The method of discourse analysis is applied here. The fact that some concepts are not used at all and others seem to find no appeal says nothing about their correctness.[8] At the climate negotiations in recent years, a profound change of perspective has taken place. It was analysed so by Rothe: 'This development is reflected in the political genesis of the UN Framework Convention on Climate Change. (. . .) The conservation of natural systems for future generations was the main goal, as it was formulated at the UN Conference on Environment and Development. In marked contrast, the support for measures of adaptation was seen as a form of resignation (. . .).'[9] In the early 1990s, there were two roles: the *polluters* (industrialised countries) and the *non-polluters* (developing countries).[10] At this time, only a small group of the negotiating nations, namely the small island countries, saw themselves as victims of global warming. Future generations were considered the main victims. Within this framework, most developing countries found themselves at the Rio Conference in 1992 far from being affected from climate change issues. To provide for future generations was a nice idea, but mitigating poverty and fostering GDP growth in one's own country took precedence for them. This self-assessment as 'non-affected parties' changed during the next decade as the quality of scientific predictions about the *regional* impacts of climate change was further improved. The developing countries realised that they (will) bear the brunt of climate change. Rothe summarises the consequences: 'It turned out that developing countries began to share the notion of a common destiny in the face of climate change: They saw themselves no longer as *non-polluters*, but as *victims*. This new collective identity of the developing countries also led to a change in their sphere of interests. They drew strength for their argument for compensation and assistance in the adjustment process out of their new self-image as victims.'[11] In terms of the four concepts of justice, this is equivalent to an extension of the concept of historical justice. This was first evidenced in the proposal introduced by Brazil during the Kyoto negotiations in 1997 to include accumulated historic carbon dioxide emissions starting in 1840.[12] At the failed Copenhagen negotiations in 2009 and at the partially successful negotiations in Cancun 2010, Durban 2011 and Doha 2012, the self-image of the South as victims[13] in conjunction with the issue of financial compensation came to the fore. One instance: During the crucial night of Copenhagen, the Chinese chief negotiator He Yafei was urged to agree to binding climate protection targets.

At which point he answered: 'Over the last hundred years, the industrialised countries have caused 80 per cent of all greenhouse gas emissions. Those who caused this problem, are responsible for the catastrophe we are now facing.'[14]

What is the ethicist's view on this? It is not ethically legitimate to consider historical justice exclusively, which was the political bargaining position of an almost closed phalanx of African, South American and Asian countries.

12

WHAT IS JUST WITH REGARD TO CLIMATE CHANGE?

The bottom line is that:

- First, there is a lot to be said for heeding the precepts of intergenerational justice as presented above. This notion of a safe emissions budget is a restriction for the current generations' consumption of atmospheric resources and it limits the leeway for distribution schemes in the present, e.g., between North and South.
- Second, the cumulative emissions between 1850 and 1990 should not be taken into account for the purpose of considerations of *justice*; compensation can (and should) ensue on the grounds of supererogatory duties like solidarity or charity or benevolence.
- Third, for the sake of *justice*, countries are accountable for their entire emissions since 1990. This should occur by means of overshooting-rights for a limited time granted to the South from the North. By contrast, to give the developing countries the choice between overshooting and a financial compensation would not be just, because in the recipe book for 'well-being', greenhouse gases are not the only ingredient. It is fair that developing countries

receive the emissions rights for greenhouse gases – they must add the other ingredients themselves.

- Fourth, for *direct* damages caused by greenhouse gas emissions between 1990 and the present (the depletion of the absorptive ability of the atmosphere is not meant here), the 'polluter pays principle' applies. All damages to the South that came to pass as a result of the actions of the North are to be compensated to the best of one's knowledge and belief. The emerging transfers from the North to the South should ideally be raised by taxes that impacts citizens with higher emissions more so than those with lower emissions.
- And lastly, the rising population of the South should be taken into account in a climate treaty. 1990 should be selected as the year of reference, as it is the year from which climate change – and therewith the contribution of population growth to climate change as well – became known to the global public. The role population growth played in climate change, however, was a taboo subject in the climate negotiations in Copenhagen and Cancun. As the UNFPA states: 'Indeed, fear of appearing supportive of population control has until recently held back any mention of "population" in the climate debate'.[1]

We don't claim that this answer to the question 'What is just with regard to the climate?' is the only possible one. Long (2011) pointed out that disagreement in climate ethics can be reasonable (not founded in error, but in difference) if the empirical evidence bearing on a case is conflicting and complex, and thus hard to assess. This is certainly true for the association between historical emissions and current well-being. However, we do assert that our account of climate justice is comprehensive and without self-contradiction.

In light of the relative loss of importance of the concept of intergenerational equity in the international arena, the question arises as to how the interests of future generations can be effectively protected. The effects of climate change pose a serious threat to the needs and interests of those living today and in the future, and this means that we must call for reforms beginning with existing political institutions. There is a growing consensus that the existing institutions are inadequate to prevent climate change (and other hazards of future generations).[2] The old division of powers in legislative, executive and judicative as conceived by Montesquieu is out of date. A new institutional framework that incorporates the interests of future generations by a 'future branch' into the current decision-making processes is necessary in creating sustainable political systems.[3] This also applies to the established Western democracies: Their development into four-branches political systems is a Herculean task that has nevertheless to be dealt with in the coming decades, because time is getting short. So far, however, not only the practical reforms, but also the theoretical explanatory approaches are inadequate. Neither liberal nor pluralistic nor participatory democratic theory renders a systematic consideration of the interests of future generations possible. Therefore, there is not only an implementation but also a theoretical deficit. In this respect, new challenges and interesting times are in store for political philosophy.

13

RIGHTS IN THE CONTEXT OF CLIMATE CHANGE: A CONTESTED TERRAIN

Rights are entitlements (not) to perform certain actions, or entitlements that others (not) perform certain actions. This section assesses whether the human rights approach adds value to efforts to address climate change to deliver justice for present and future generations. Can – and should – the language of justice be replaced by the language of rights?

'Rights talk' is used in several ways in the context of climate change. First, it is argued that presently living people in developing countries should have a right to greenhouse gas emissions if these emissions are indispensable for a minimum standard of living (so called 'subsistence emissions'[1]). We call this approach the 'emission rights for present people in developing countries' approach (in short: 'poor present people's emissions rights-approach').

Second, it is argued that the postulation of 'rights' can help prevent dangerous climate change. Human rights, so the argument goes, provide an effective clarion call for stronger mitigation efforts.[2] Granting such negative claim-rights[3] to people would impose duties on present people or mankind in general to refrain from emitting too many greenhouse gases. This argument takes two variants:

1a) Proclaiming *present* people's human rights will help stopping the present emitters from emitting excessive greenhouse gases (in short: 'present people's right to a stable climate-approach').

1b) Proclaiming *future* people's human rights will help to stop the present emitters from emitting (in short: 'future people's right to a stable climate-approach').

Let's start with the question whether or not there should be a 'right to emit'. The chapter *International justice in a world with population growth* (e.g. Figure 9.4) has showed that a growing population leads to shrinking per capita allocations of a scarce resource. Prima facie, this seems to show that there cannot be inalienable rights to greenhouse gas emissions. With a growing population, the ration of a scarce resource per head must be constantly recalculated. In the debate about the so called repugnant conclusion[4] philosophers make the assumption of a quickly growing population and ponder the ethical consequences of this scenario. One result is that the happiness or utility of each individual will asymptotically reach a very small number (but not zero) which symbolises the survival minimum. The same reduction happens with the allocation of greenhouse gas emissions, provided the worldwide population as a whole stays within the safe emission budget and does not eat up the atmosphere's absorptive capacity that was allocated to future generations. An absolute right to the atmosphere in general, as proclaimed by Shue, Vanderheiden or Lumer however, is not amenable to the continually fluctuating amount of emissions to which each person is entitled.[5]

But what would a lawyer make out of all this? He would litigate and if a codified 'right to emit' existed he would win in court. Ronald Dworkin has coined the cogent metaphor 'rights as trumps'. He explains their relationship to utilitarianism as follows: 'Rights are best understood as trumps over some background justification for political decisions that states a goal for the community as a whole. (...) I shall assume that the background justification with which we are concerned is some form of utilitarianism which takes, as the goal of politics, the fulfilment of as many of people's goals for their own lives as possible.'[6] In utilitarianism, the criterion of utility determines whether an action is right or wrong. For the utilitarian approach, the offsetting of the change in utility for various options is of key importance for the correct course of action. Thus, for example, it would be justified to torture a kidnapper if in so doing one could rescue the lives of several children. A captured jumbo jet with 200 people on board could be shot down if it were flying towards a skyscraper with 2,000 people inside.[7] Rights as trumps overrule this calculus. If one has a right not to be tortured, then the government cannot torture a person under any circumstances. If one has the right to emit greenhouse gases, then the government cannot hinder her under any circumstances. 'Rights talk' renders all the ethical considerations so far displayed in Part II of this book to a certain extent useless. It's the law, stupid! You have to obey it! With rights-based approaches, actions are wrong if they violate rights. But who is to decide whether there are rights in the first place? What are rights? What is the function of rights? How can rights be justified? What is the relationship between ethical norms, moral rights and legal rights? What do the answers to these questions mean for a proclaimed 'right to emit of present poor people'?

According to Feinberg, 'to have a right is to have a claim *to* something and *against* someone, the recognition of which is called for by legal rules or, in the case of moral rights, by the principles of an enlightened conscience.'[8] Today's widespread concept of rights as claim-rights goes back to Hohfeld (1919), for whom it was only one of three concepts of rights, however. The legal theorist Jeremy Waldron writes in his book *Theories of Rights*: 'Hohfeld's claim-right is generally regarded as coming closest to capturing the concept of individual rights used in political morality.'[9]

Bentham has put forward the opinion that real (or enforceable) rights come from real (or legislated) law, recognisable by the duties imposed on others, not by normative contents of aspirational documents. He writes: 'On a natural right who has any idea? I for my part have none: a natural right is a round square or an incorporeal body. What a legal right is I know. I know it was made. I know what it means when made. To me a right and a legal right are the same thing, for I know no other. Right and law are correlative terms: as much so as son and father. Right is with me the child of law (. . .).'[10] The legal and the ethical discourse overlap, but they should be distinguished. *Moral* rights are special

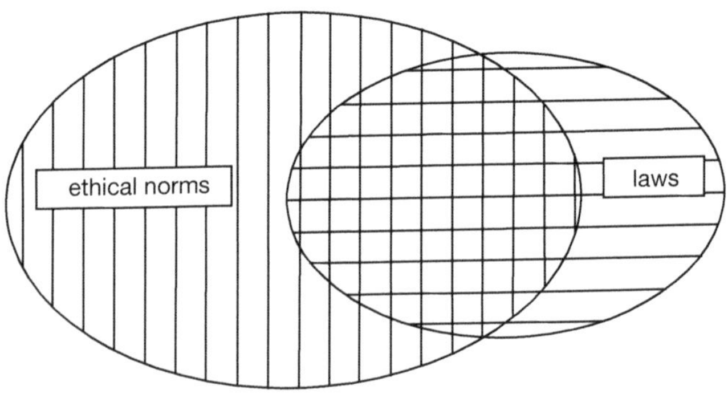

Figure 13.1 The relationship between ethical norms and laws
Source: Tremmel (2009, 48).

cases because they are part of the ethical discourse, but they already build a bridge to laws. Let's ignore moral rights for the moment. Then the relationship between laws and ethical norms can be depicted by two intersecting circles.

Three cases are possible.

Firstly, there are ethical norms ('One should...') that are not embedded in positive law – some will be non-contextualised, others could be codified into legal terms, but the political majority is reluctant to do so. Norms that have not (yet) been codified are depicted by the non-overlapping part of the 'ethical norms circle'.

Secondly, there is the group of ethical norms that are at the same time laws and vice versa (intersecting part of the two circles).

The third case (non-overlapping part of the 'law' circle) refers to legal norms that are not ethical. For example, the Nuremberg Racial Laws of Hitler's 'Third Reich' were blatantly unethical. Nevertheless, they were codified in positive law. Another example (still not comparable to the Nuremberg Racial Laws, but bad enough) are the apartheid laws in South Africa until 1994. They made apartheid legal, but not moral.

Unlike Bentham, most ethicists nowadays employ the term 'rights' in the ethical sphere ('moral rights') and the legal sphere ('legal rights'). What is the relationship between these two? Taking a bird's-eye view, codified law is usually adjusted according to the changes in the moral convictions of a society, sooner or later. If there were a consensus that poor present people had a moral right to greenhouse gas emissions, it would only be a matter of time before these rights were enshrined in international law. Ethics precedes law-making (or, at least, it should). Thus, we can start with discussing the validity of moral rights. So what are moral rights actually? What is their origin, their purpose and their justification?

The language of the human rights discourse was coined during the Age of Enlightenment.[11] In her path breaking book *Inventing Human Rights,* the historian Lynn Hunt argues that the idea of unalienable rights only became convincing to political philosophers and theorists in the eighteenth century. It must be distinguished from older ethical concepts, e.g. the notion of *justice*. The great Greek philosophers never spoke explicitly of individual or human rights. The concept of human rights also does not appear explicitly either in the Hebrew Bible or the New Testament. The duties that these documents mention are owed to God.

Leif Wenar, author of the entry 'Rights' in the *Standford Encyclopedia of Philosophy*, argues that rights are in fact much older than the age of enlightenment: 'Intellectual historians have tangled over the origins of rights. (...) Yet insofar as it is really the emergence of the *concept* of a right that is at issue, the answer lies beyond the competence of the intellectual historian and within the domain of the anthropologist. Even the most primitive social order must include rules specifying that certain individuals or groups have special permission to perform certain actions. Moreover, even the most rudimentary human communities must have rules specifying that some are entitled to tell others what they must do. Such rules ascribe rights.'[12]

Well, during the time of slavery, there was a right called 'the right of the first night'. According to this provision, when two slaves married, the slaveholder had the right to spend the first night with the virgin wife. This right was generally accepted; the newly married husband had no choice but to respect this right as well. But it was profoundly immoral.

Prerogatives of monarchs and power-wielding classes existed since ancient times. But these 'rights' (in Wenar's terminology) must be clearly distinguished from the concept of universal human rights. Human rights refer to those rights persons have qua human beings.[13] Entitlements classified as 'human rights' represent that to which each and every individual is entitled, without discrimination. These human rights took centre stage only in the seventeenth and eighteenth centuries, in the writings of Thomas Hobbes, John Locke or Jean-Jacques Rousseau. The idea of human rights gained broad acceptance during the eighteenth century, when the *American Declaration of Independence* (1776) proclaimed that all men are endowed by their Creator with certain unalienable rights, among these life, liberty and the pursuit of happiness. A few years later, the French revolutionists drew up the *Declaration of the Rights of Man and Citizen* (1789). According to Hunt, human rights were grounded in the rejection of torture as a means of finding the truth; the changing idea of human relationship displayed by novelists, playwrights, and artists; and the spread of empathy beyond insular communities. Before human rights were codified in corresponding declarations, enlighteners had intensively debated their existence. Thus, the proclamation of 'moral rights' preceded the creation of legal rights.[14] But the landslide victory of the human rights concept has not gone unchallenged. It was contested from the very beginning. The essayist and member of the British parliament, Edmund Burke, agreed with the *purpose* of the human rights talk during these days, namely to protect the individual from all political games and governments.[15] But Burke puts his finger on the crucial question of *justification* when he writes in his pamphlet *Reflections on the Revolution in France* (1790): 'We know that we have made no discoveries, and think that no discoveries are to be made, in morality [. . .]'[16] The French enlightener Marquis de Condorcet holds the opposite point of view: '[T]hese rights that are at once so sacred and so long forgotten.'[17] Now, was the concept of rights invented or discovered? The question of whether people have rights is not comparable to the question of whether people have noses, because rights are abstract matters.[18] If human rights are self-evident, as claimed in the documents of 1776 or 1789, then why does this assertion have to be made at all, and why was it made only under certain conditions, thousands of years after philosophers had started debating moral questions? Human rights did not fall out of the clear blue sky; they are neither God-given nor a product of nature. Rather, they are an invention of man. Regarding the right to life, this innovation entailed first converting the sentence 'people should unconditionally be protected from being arbitrarily murdered' into a moral right and then replacing this moral right with the legal right 'Every person has a human right not to be arbitrarily deprived of his life' (International Covenant on Civil and Political Rights, 1976, article 6.1, and the predecessors to this right in various human rights declarations and constitutions since the eighteenth century). But the precondition for the moral and for the ensuing legal right – for rights talk in general – is that ethical theories (deontological or consequential) arrived at this very conclusion. Rights are not discovered. They are moral innovations that must ethically be justified instead being posited. There *are* thus innovations in the realm of morality, and human rights were a particularly successful one. The language of rights is a powerful instrument in moral and political debates that trumps all other language games.

We agree with Wenar when he writes 'Moral rights are grounded in moral reasons.' but we disagree with him when he writes 'Natural rights are the sub-class of moral rights that humans have because of their nature.'[19] 'Nature' cannot be a justification for human rights. The biological nature of human beings or the natural system does not tell us what is right or wrong. The term 'natural rights' should be avoided. Man himself must decide what is rightful.

The human rights perspective is one, but certainly not the only possible *point of origin* for theories of morality and theories of justice in particular. Some scholars doubt that the idea of human rights is a valid *starting point* for political morality in general and theories of justice in particular. That is emphasised by contemporary critics of the moral rights narrative. The redundancy theory of rights, for example, explores whether the semantic content of the language of rights could be reproduced in its entirety with the language of duties.[20] There are other schools of thought as well: 'Even in the liberal tradition, some philosophers insisted that rights could be taken seriously only if they were understood to be based on a prior theory of social and political morality such as the theory of utilitarianism', writes Jeremy Waldron.[21] Annette Baier states: 'We do of course have legal rights, but to see them as backed by moral rights is to commit oneself to a particular version of the moral enterprise that may not be the best version. As Hegel and Marx pointed out, the language of rights commits us to questionable assumptions concerning the relation of the individual to the community, and, as utilitarians have also pointed out, it also commits us more than may be realistic or wise to fixing the details of our moral priorities in advance of relevant knowledge that only history can provide.'[22]

With rights-based approaches, actions are wrong if they violate rights. But often the very existence of certain rights is contested. Just think of the lengthy dispute on the priority of either political and civil rights or socioeconomic rights between capitalist and socialist countries (both conceptions of rights are enshrined in the *Universal Declaration of Human Rights*, published by the United Nations in 1948). In Jefferson's days, what we call now 'unalienable rights' did not apply to those without property, slaves, free blacks, a number of religious minorities, and women.[23]

Later, a majority attributed moral rights to women. By this shift in consciousness, women 'received' these rights. The underlying conditions had not changed, but according to the collective social awareness, these 'rights' suddenly existed. The attribution of rights, and thus their existence, depends on emotions as much as on reason.[24] The claim of self-evidence of a human right strikes a chord if we feel horrified by the violation of the right in question. Up to today, it all depends on determining what is 'no longer acceptable' for a majority of people.[25] Could a general right to a quite high minimum income soon see the light of day?[26] This right is already formulated as a moral right in many European countries. The goal of its proponents is to translate it into a legal right, that is, a law, as soon as possible because only then will there actually be a change in the income distribution of a society. By speaking in terms of 'rights', a moral claim receives precedence over other ought-statements. He who demands that the rest of society recognise a right wishes to change the system of morals, because if he succeeds in instilling said moral right into the consciousness of the general population, then this right exists henceforth – and therewith a new group of rights holders. Thus, whether or not a right is established depends on whether or a not a given circumstance is an entirely unacceptable notion for many people. This can be anywhere on the spectrum from a society unprotected from torture by its government or a society with an inadequate minimum income. The result is that the discourse on the creation (and theoretically also the abolition) of rights continues perpetually. Hunt writes: 'Rights remain open to question because our sense of who has rights and what those rights are constantly changes. The human rights revolution is by definition ongoing.'[27] But people's emotions might move forward *or* backward, so we cannot be sure that ever more rights will be established. As we have seen, rights are not self-evident, but a matter of convention.

Apart from empathy, are there any logical criteria for ascribing rights? It is commonly believed that an individual must have interests before he or she can become a rights-bearer.[28] The basis for interests are needs. That is why we can ascribe rights to animals or even aliens, but not to rocks or to artefacts like the Taj Mahal. Interests must somehow be compounded out of conations. A rights-holder must be capable of being a beneficiary in his own person, and an object without needs is incapable of being harmed or benefitted. This is an important logical criterion for ascribing rights.

All of these criteria apply to the 'poor present people emission rights' approach. Thus there are no principle or logical hindrances to postulating a human right to GHG emissions. Is it, however, advisable to employ the trump? Should the claim to 'subsistence emissions' have precedence over other ought-statements? It is advisable to go forward with caution when answering these questions: With a growing population, the ration of a scarce resource per head must be constantly recalculated. A proclaimed absolute *right* to greenhouse emissions of a fixed amount, say 2 t/head, is not amenable to the continually fluctuating amount of emissions to which each person is entitled. If we codified such a right, total GHG emissions might rise even above 560 Gt of greenhouse gases until 2050. This would not be in the interest of future generations who will suffer in the future from the impacts of dangerous climate change. And it is not in the interest of the present people who already suffer (or will suffer in the near future) from climate change. This brings us to the first variant of the 'right to a stable climate-approach'.[29]

14

THE 'PRESENT PEOPLE'S RIGHT TO A STABLE CLIMATE' APPROACH

Does human rights provide an effective clarion call for stronger mitigation efforts? Indeed, such a clarion call was part of the motivation behind the 2007 *Male' Declaration on the Human Dimension of Climate Change* in which a group of the world's most vulnerable small and developing states, for the first time in an international agreement, stated that climate change had 'clear and immediate implications for the full enjoyment of human rights.'[1] These states were frustrated at the slow pace of progress in the UNFCCC negotiations and wished to focus attention on the victims of climate change. The Male' Declaration led in turn to the Human Rights Council tasking the Office of the UN High Commissioner for Human Rights (OHCHR) to conduct a study on the effects of climate change on the full enjoyment of human rights. This study concluded that climate change did have 'a range of implications for the effective enjoyment of human rights' including 'direct' effects such as extreme weather events posing a threat to the right to life, but also 'indirect and gradual' effects on human rights such as increasing stress on health systems.[2]

The OHCHR Report recognised existing human rights instruments contain rights which would be breached by the failure of governments to address climate change. These include the right to a standard of living adequate to health and well-being (*1948 Universal Declaration of Human Rights*, article 25) and the right to the highest attainable standard of health (*1976 International Covenant on Economic, Social and Cultural Rights (ICESCR)*, article 12 (1) and (2) (b)), the human right to life (*1966 International Covenant on Civil and Political Rights (ICCPR),* article 6 (1) and the human right to subsistence (*ICESCR,* article 11). These rights would be violated by governments' failure to limit dangerous climate change, given scientists' predictions of increased exposure to tropical diseases and deaths from extreme weather events, as described in the Part I of this book.

The OHCHR Report stated that individual claims for breaches of human rights obligations were problematic owing to the difficulty of attributing climate change impacts to individual states.[3]

In 2005, the Inuit people brought a case before the Inter-American Commission on Human Rights in which they claimed that the United States by failing to curb its GHG emissions had violated their fundamental human rights.[4] The Commission dismissed the petition without giving reasons.

It seems that the 'present peoples' right to a stable climate-approach' has some problems to convince judges around the globe. Let's assume for a moment that the juridical difficulties for a *legal* right to a stable climate were insurmountable. Could then nevertheless a '*moral* right to stable climate' be postulated? This a difficult question. On one hand, rights talk should not raise false hopes. For the newly created group of holders of this moral right to a stable climate it would seem as a insidious mockery if their new instrument proves to be useless in court permanently or for a long time to come. On the other hand is there a manifesto function of 'moral rights'. Birnbacher explains: 'A second function of the language of rights is the appeal to rights that are not, or not yet, part of the respective normative system but are postulated as necessary or desirable additions by moral or political reformers. This is the *manifesto* use of the language of rights or, as it may be termed, its *revisionary* use. In this use, rights are postulated in the knowledge that they are not as a matter of fact recognized, or only in special cases or by very few communities, with the hope that they will come to be recognized more widely at some future point of time.'[5] There are many examples of cases in which we speak of rights even if these purported rights cannot be enforced for a long time. Article 1 of the *Universal Declaration of Human Rights* (1948) starts as follows: 'All human beings are born free and equal in dignity and rights.' That is, of course, counterfactual. There is still a long way to go before we all live with the same dignity. Article 24 reads: 'Everyone has the right to a standard of living adequate for the health and wellbeing of himself and of his family, including food, clothing, housing and medical care and necessary social services, and the right to security in the event of unemployment, sickness, disability, widowhood, old age or other lack of livelihood in circumstances beyond his control.' Even in developed countries, there are many homeless people, people without health insurance, or without insurance against unemployment.

At any rate, it is worthwhile to ponder the difficulties of the '*legal* right to a stable climate-approach' *before* one decided to postulate a 'moral right to a stable climate'. Now, what are these juridical difficulties?

Firstly, as already discussed in the context of the polluter pays principle above, there is the difficulty of causation. Floods, draughts and thunderstorms may have a combination of causes.

Secondly, very often the complainant' government will not be responsible for the harm done. 'For example it is not much use for a citizen of Kiribati to bring a human rights related claim against their own government in relation to climate change damage' writes Peter Lawrence.[6] In the above quotation, Lawrence is touching upon the thorny issue of human rights complaints brought against governments who are responsible for causing a particular harm that does not cause a violation of human rights 'internally', but 'externally' (outside the jurisdiction of the state responsible for the harm). International law in this area of 'extra-territorial harms' remains contested.[7]

Third, greenhouse gas emissions are the aggregate result of individual activities and the industrial activities of corporations and states. A plethora of actors are involved. But the obligation to ensure human rights is typically regarded as the responsibility of national governments only. While there have been individual instances in which international corporations were held responsible for breaches of human rights, these have been determined on a case-by-case basis and have not been incorporated into any treaties or legal documents.[8]

Individual human rights litigation has necessarily a reactive quality, with the requirement that the damage must already have occurred, whereas what is needed for effective climate change policy is preventive action.

This brings us to our second variant of the 'right to a stable climate-approach', the 'future people rights' approach. But can future people have rights at all?

15

THE 'FUTURE PEOPLE'S RIGHT TO A STABLE CLIMATE' APPROACH

The third part of Partridge's anthology *Responsibilities to Future Generations* (1980c) discusses whether future generations can be said to have rights.[1] The editor puts paramount importance on this question:

> Thus, if future generations have rights-claims against us, they will have no cause to be 'grateful' to us for preserving a viable ecosystem; for they will have received their due. On the other hand, if we violate this duty, their appropriate response will be not simply *regret* but moral *indignation.* Moral duties born of *rights* weigh more heavily upon the duty-bearers. Thus, to the degree that our policy-makers and legislators respond to valid moral arguments, the interests of future generations will be far better served if we can succeed in defending the notion that succeeding generations have rights-claims against the living who, in turn, have the moral duty to respect and respond to these rights.[2]

Needless to say, future people are not able to renounce their rights in the present. But the same applies to many contemporaries, and that does not mean they have no rights. Babies or people in a coma cannot waive their rights, but according to a broad consensus they still have rights, e.g., the right to live. If a starving person is too weak to express himself, he has by no means forfeited his right to be fed. 'The fact that future subjects are not able to assert any rights they may have towards the present generation for logical reasons whereas contemporary subjects may not be able to do so for contingent reasons, cannot reasonably justify denying moral rights from to former, but not to the latter', writes Birnbacher.[3] Thus, it would be wrong to assume that the ability to waive a right is a precondition for ascribing a right. Nor is the ability to understand what a right is and to set legal machinery in motion by one's own initiative a prerequisite for having rights.[4] For instance, the rights of infants are normally claimed and defended by appointed counsels or public agencies. Therefore, the (future) rights of persons who have not yet been born can be represented by proxies or attorneys empowered to speak in their names.

But many writers have denied that future people can have moral rights.[5] Winfred Beckerman has long been considered the most renowned critic of all concepts based on 'rights' of future generations.[6] Over the past years, he has also criticised theories on generational justice. Those are two different things. Even if one could not speak of 'rights of future generations' in a meaningful way, a meaningful theory on generational justice would be conceivable. Beckerman sums up his theses as follows:

> My argument is really very simple and can be summarized in the following syllogism:
>
> (1) Future generations – of unborn people – cannot be said to have any rights.

(2) Any coherent theory of justice implies conferring rights on people.

Therefore, (3) the interests of future generations cannot be protected or promoted within the framework of any theory of justice.[7]

Let us first examine statement 1: The conceived but unborn child has the legal capacity to hold rights, for instance the right not to be killed if the conditions for a legal abortion are not fulfilled. But below we will exclusively deal with non fathered, 'potential' individuals. According to Beckerman, the general proposition that future generations cannot have anything, including rights, follows from the meaning of the present tense of the verb 'to have'. 'Unborn people simply cannot *have* anything. They cannot have two legs or long hair or a taste for Mozart', Beckerman writes in the *Intergenerational Justice Review*.[8] Beckerman's argument is correct, but of minor importance. It reminds us that we should use future tense instead of present tense, that is, to say: 'future generations will have rights' instead of 'future generations have rights'. We normally use future tense when we refer to characteristics and attributes of future people, their rights, or even their noses, and rightly so. But Beckerman's argument cannot be used to denounce the term 'rights', or to replace 'rights' by 'just claims', 'needs', 'interests', 'wishes', or the like. If future generations do not have 'rights', they do not have 'interests' and so on, either. They *will* have interests, just as they will have rights. If we want to favour the term 'interests' over 'rights', we must find other arguments. The hint that we should use the future tense instead of the present tense is just a minor aspect that has been misunderstood in the literature as a substantial issue. In fact, Beckerman's argument is insubstantial and would only have required some semantic clarification.[9]

His question 'Do future people have rights?' only needs to be rephrased to 'Will future people have moral rights?', the headline of the next chapter.

Will future people have moral rights?

The substantial, correctly formulated questions are: 'Will future people have moral rights?' or 'Can future people correctly be said to have rights in the future?' Now, these questions cannot be answered in the negative as easy as Beckerman wishes. At the contrary, these questions are easy to answer, prima facie: if present people have human rights during their lifetime, so will future people. Annette Baier puts it this way: 'No one doubts that future generations, once they are present and actual, will have rights, if any of us have rights.'[10]

For clarity we want to emphasise that we discuss universal human rights in the following. These 'general rights' 'are rights that people have in virtue of their humanity, and not because of the nation or state into which they were born or any actions that they have performed.'[11] We neither discuss any prerogatives nor 'special' rights like the right to payment after delivery of a good in virtue of a valid contract.[12]

Do we have present obligations to people who will exist in the future?

It is almost undisputed that we have present obligations and responsibilities towards future persons, even if their identities are 'not-yet-determined'. Even critics of the rights of future persons like Beckerman or De George freely admit that we have obligations towards them. A great majority of philosophers supports a comparative standard with regard to intergenerational justice; that is a standard that determines the wellbeing of future generations by comparing it to that of today or to that of earlier generations. Within this group, a considerable number of scholars postulate principles of strict equality between generations ('as good as'). A proponent of such an egalitarian standard of intergenerational justice is Brian Barry who demands 'that the overall range of opportunities open to successive generations should not be narrowed.'[13] De George formulates the egalitarian maxim negatively. With regard to family generations, he says: 'Parents do not owe their children better lives than they had.'[14] An even greater number of texts in literature on intergenerational justice use the formulation 'at least as good'. See some examples: just like John Locke, 300 years ago ('at least as much and as good'),[15] the philosopher Gregory Kavka suggests: '[. . .] I interpret this to mean that, in this context, the generation in question leaves the next generation *at least* [emphasis added] as well off, with respect to usable resources, as it was left by its ancestors'.[16] Rakowski puts it this way: 'Everyone born into a society is entitled, as a minimum [emphasis added], to the same quantity of resources that all who participated in the original division of the community's goods and land received.'[17] Likewise, Dieter Birnbacher argues: 'Everyone should leave *at least* [emphasis added] as many natural resources as it was left to him. What someone has inherited, he should pass on undiminished ("to sustain"), and *possibly* [emphasis added] increased ("to cultivate"), to future people, be it as a private citizen or as a representative of a collective.'[18] James Woodward adds 'opportunities', but he sings the same tune: 'Each generation ought to leave for succeeding generations a total range of resources and opportunities which are *at least* equal [emphasis added] to its own range of resources and opportunities.'[19] But the idea of an obligation to *improve* the quality of life for future generations is also expressed sometimes. Karl Marx writes in the third volume of *The*

Capital: "Even a whole society, a nation, or indeed all concurrent societies taken together are not the owners of the earth. They are only its possessors, its beneficiaries, and like boni patres familias, they must hand it down to succeeding generations in an *improved* condition [emphasis added]."[20]

And the philosopher John Passmore claims that: "(...) [W]e ought to try to improve the world so that we shall be able to hand it over to our immediate successors in a *better* [emphasis added] condition, and that is all."[21]

The principles laid out above are conflicting in terms of the extent of our obligations to future generations; however, all acknowledge that obligations do indeed exist. None of these conceptions of intergenerational justice purport that we have no obligations to coming generations. This does not come as a surprise. *Prima facie*, in the universalistic tradition that coins our moral reasoning since the Enlightenment, differences in time should matter no more than differences in race or gender. It seems to be both logical and intuitive to extend the principles concerning our dealings with our contemporaries to our dealings with future people. Once a moral maxim of impartiality is accepted as a core of moral reasoning, it seems plausible to extend the realm of moral patients to future persons. The *moral* point of view is a point of view beyond all particular perspectives, and any attempt to privilege the present over the future is prima facie at odds with impartiality. But there are counterarguments against the notion that we have obligations to future generations. These counterarguments will be dismissed one by one in the following.

Persons with indeterminate identities

In many respects, the distance in time resembles distance in space. Many ethically minded people feel obliged to relieve the suffering and despair of people with indeterminate on the other side of the earth. The enormous outpouring of aid in Europe and the USA after the tsunami on 26 December 2004 with its roughly 230,000 victims is proof of this long-distance ethics. 'Location in space is not a morally relevant feature of a person, determining his worthiness for considerations or aid. Why should location in time be any different?' asks Kavka.[22] In the case of future persons, indeterminacy is a result of non-actuality. In the tsunami case, we donate to people who are indeterminate to us because we believe that the agencies know best who needs the money most urgently. We often feel obligated towards indeterminate people, and it is irrelevant whether they are already born or not. I should refrain from leaving broken glass on the beach, not for the 'sake' of a particular beneficiary of that duty, but to prevent possible harm to anyone.[23] Or, if I dig a mine shaft in a remote hiking area, I am obliged to cover it, so no one gets hurt.[24] In the same way, I am obliged to leave a very remote campsite neat and tidy when I leave.[25] And that has nothing to do with whether or not the person who might get harmed if I do not meet my obligations has already been born. If we take the moral point of view, we should care for the well-being of present *and* future individuals, independently of their identities. Muñiz-Fraticelli adds an important point: 'Most laws, for instance, are not written with specific, identifiable individuals in mind but rather in generic language which is not identity dependent. We may not know whether Anna or Ben has bought the house on the corner, but we can be sure that whoever is now the owner is equally obligated to pay taxes on the property. Why then, to contemplate a harm to a future person, must we identify the particular individual who will actually exist in the future and not merely point out that [. . .] the category of "future person" will not be an empty set?'[26] The same applies to case law: when a case is before a court, the judgement is passed to the individual, but it affects the way everybody else in the same situation would also be treated.[27] And Partridge convincingly argues that moral principles apply to individuals by description and not denotatively; that is, due to shared general qualities and relations rather than qualities that distinguish persons as individuals like their genetic codes or personalities.[28]

To cut a long story short: the indeterminacy argument does not relieve us from our obligation to take the interests and rights of future, yet-to-be-determined persons into account in our present actions.

An unknown number of future persons

In order to assess our obligations towards future persons, we must make some assumptions about the number of future persons. How do we know if there will be any at all?

The convincing rejoinder is: probability. Excellent scientists have made state-of-the-art forecasts regarding the global population. Given the current global population growth rate, we can safely assume that there will be many people on earth over the next centuries. According to the medium scenario of the United Nations' long-range population projections, the world population will rise from 6.6 billion in 2007 to a maximum of 9.2 billion in 2075 and then decline to 8.3 billion in 2175. The return to replacement level fertility coupled with increasing longevity in the medium scenario will produce a steadily increasing population after 2175 that will reach nine billion by 2300.[29] According to demographers, that is the most likely scenario. Demographers are still discussing whether there might be 8, 9, or 10 billion, but none of them assigns any probability to a scenario in which there will be zero people on earth in the year 2300. Let us take nine billion as a starting point, since that is the best prognosis of the scientific community at the moment. Provided we accept that these people, once they exist, will *then* be rights-bearers, we are justified to speak of obligations of the current generation. The event that could wipe out human beings from the face of the earth to the greatest degree is a nuclear catastrophe. But that (or other man-made catastrophes) can scarcely release us from our responsibility towards future generations, because it is up to us to avoid it.[30] We would only be relieved from our obligations if we could assume that an act of *force supérieure* will eliminate mankind (for example a cataclysmal meteorite). But up to now, no such meteorite impact has ever taken place in the history of mankind.

Contingency, non-actuality and insecurity of needs

Another objection against ascribing rights to future people might be that future persons are contingent on our actions. But this objection – the 'non-identity paradox' – has already been refuted.[31] Another objection might be that what we do for future generations is subject to greater uncertainties than aid granted to remote contemporaries, because we cannot foresee the exact needs and preferences of future people.[32] It is of key importance not to confuse needs and wants here. Wants and preferences vary from culture to culture and from individual to individual. Fashionable subjectivist and cultural relativist approaches thus relate to wishes, wants, or aspirations rather than needs. The needs of every member of every generation are identical, no matter which age or culture she lived (lives, will live) in. Most probably, future individuals will also need air to breathe and water to drink.[33] Therefore, the argument that we have no obligations towards future generations because we cannot know all their higher preferences loses ground. Partridge writes: 'The very enormity of the changes that are projected, or imminent, may render a finely tuned science of forecasting somewhat irrelevant. For whatever their tastes in music or poetry, or whatever their preferences in sports and other amusements, our descendants will need croplands and watersheds to supply their food and water.'[34] Thus, the insecurity objection is not very powerful either.

What about the non-actuality of future people? Imagine a freshly married couple, both of them physically able to get children, discussing whether they should stop using contraceptives in order to have a baby. It would be wrong to say that they have no obligations for the child at present. The fact that they have not picked a name for the baby yet does not affect their duties and responsibilities. Supposing the husband has travelled around the world and enjoyed himself instead of finding a job and earning money. His wife could rightly ask him to take measures to support his family when she is pregnant or on maternity leave. The unborn child will have needs, e.g., food and shelter, and it will have corresponding rights, like the right to life. The parents are obliged to make sure *now* that it has a home and can be fed after birth.

Claiming that future people have not yet rights does not necessarily imply that we cannot violate future individuals' rights today. That would only be the case if we conceded that present rights alone can constitute present obligations.[35] But *future* rights can also constitute *present* obligations.

The relationship of rights and obligations

It should be clear now that we have obligations towards future generations. It is also undisputed that rights are somehow related to obligations or duties.[36] There are four possible relationships between rights and obligations:

1) Whenever party A[37] has an obligation in relation to party B, B has a right in relation to A. Whenever party A has a right in relation to party B, B has an obligation in relation to A. In this conception, rights and obligation are just two sides of a coin. They are strictly correlative. If we admit that we have obligations to future generations, they will automatically have rights.
2) Whenever party A has an obligation in relation to party B, B does not necessarily have a right in relation to A. Whenever party A has a right in relation to party B, B has an obligation in relation to A.
3) Whenever party A has an obligation in relation to party B, B has a right in relation to A. Whenever party A has a right in relation to party B, B does not necessarily have an obligation in relation to A.
4) Whenever party A has an obligation in relation to party B, B does not necessarily have a right in relation to A. Whenever party A has a right in relation to party B, B does not necessarily have an obligation in relation to A.

Philosophers have mixed opinions on these definition options.[38] For Beckerman, it is not possible to deduce that all obligations imply rights from the proposition that all rights imply obligations. He does admit that we have obligations to future generations, but: 'One can think of innumerable situations in which one's behaviour will be influenced by some conception of what our moral obligations are, without necessarily believing that somebody or other must have some corresponding rights.'[39]

Consider an orphan who has, most people would intuitively say, a right to be raised in a family. But that does not imply an obligation of a certain family (or any family) to adopt him. Waldron also considers option 3 when he says: 'I can say, for example, that a child in Somalia has a right to be fed, meaning not that some determinate individual or agency has a duty to feed him [. . .]' Yet, he adds: '[. . .] but simply that I recognize his interest in being fed as an appropriate ground for the assignment and allocation of duties.'[40] For Waldron, at least in this example, there are no rights without obligations.

Obviously, the relationship is complex. The problem is that the community of philosophers has not yet agreed on a final definition of the terms 'moral right' and 'moral obligation'. So, the definition criterion 'common use in the scientific community' yields no clear result. I see no logical criteria that would make any of the four definition options more compelling than the others, so it is a matter of language convention. I personally tend to opt for a strict correlation (option 1). Most authors who cling to other options do this because of moral 'in rem' rights. Bandman explains: 'A right in personam is made against a specific or determinate person or group such as one finds in the right of a creditor against a debtor. Such rights correlate with specific duties of determinate individuals. In rem rights are not against specific nameable persons, but against the world at large. The right of an accident victim to assistance implies a duty by anyone who happens to be in a position to help is an in rem right.'[41] We would not speak of a 'right' of the accident victim in such a context. If we search for analogies in the sphere of legal rights (which we will treat in detail later), we see a clear correspondence between 'legal rights' and 'legal obligations' which makes definition 1 appear most conclusive.[42] If an orphan has a legal right to be raised in a family, then the state is obliged to make sure he or she is adopted as soon as possible. The orphan can call on the power of the state. The alleged moral rights against 'the world in general' or rights 'in rem' correspond either to legal rights against state authorities (then legal measures can be directed to their representatives) or they are no legal rights at all. In a society, people use different wordings for their moral principles, but the language of law applies to all alike. In this particular case, this seems to be an argument to choose option 1 in the definition-making process.

Parenthetically speaking, that does not mean that party A only has rights if it also has obligations. One does not have to be a duty-bearer to be a rights-bearer. Even newborn babies have rights, but they have no obligations yet.[43]

Unusual wordings

So far, I have listed a series of points that indicate that it makes sense to say: 'Future people will have moral rights. We are obliged to respect these rights, today.' Nevertheless, that leads to unusual wordings that shall now be discussed in more detail. We are used to employ 'rights talk' if there is at least a theoretical chance of enforcing the rights, be it by convincing others by means of moral arguments or by taking legal action.

Consider the following: tigers are now a threatened species. Suppose the people alive today were obliged to preserve them because future people will have a right to experience them. If the people alive today do not meet their obligation, so tigers become extinct, will future people – perhaps in 300 years from now – have the right to see tigers in *their* present? Beckerman does not think so, and his example – slightly modified – is as follows:

> In the case of rights to particular physical objects, like a right to see a live tiger, it is essential that the tiger

> exist. [...] Thus for the proposition 'X has a right to Y' to be valid, where Y refers to some tangible object, two essential conditions have to be satisfied. First, X must exist, and second, it must be possible, in principle, to provide Y. In the case of the right of a future person to see live tigers, for example, one of these two conditions is not satisfied. He exists in *his* present (in *our* future), but tigers do not exist. And before tigers became extinct, the tigers existed but the future person did not exist. Hence, insofar as it is implausible to say that the future person *had* the right to the preservation of live tigers before this future person existed it must be implausible to say that a member of a non-existent unborn generation will have any rights to inherit any particular asset.[44]

It is unusual, but is it implausible? Let's consider the nexus of human rights and climate change with one eye on Kiribati, an island nation located in the central tropical Pacific Ocean, and one of the world's most vulnerable nations in terms of the effects of climate change.[45] It is well possible that Kiribati will vanish in the next decades or centuries. Let's assume, this will happen 2200. Now, is there a right to live in their home country for future Kiribatians? Obviously, this human right *could not* be respected. Core human rights have an appeal due to their embodiment of universal ethical values. But > Ought implies can < is a fundamental rule in ethics. This is the very point that makes it difficult for many to speak of rights of future generations. Speaking of 'needs' or 'interests' of future Kiribatians would not pose a problem for many scholars in this example. But 'rights' that cannot be enforced, so lawyers cannot do anything about them, appear strange to many. Beckerman concludes that it would be wrong to speak of rights in such a context.

It should be noted that demise of Kiribati lays in the hands of the present generation and is still avoidable *at present time.* An interest or potential right of future Kiribatians to life in Kiribati in 2200 differs in this regard from an interest of future Kiribatians to be contemporaries of Shakespeare in 2200. The very function of rights talk is to make the present generation act. We do not know in 2012 if those Kiribatians alive in 2200 will have the right to life in their home country (because if their island will be lost forever, there will not be such a right in 2200. But if we act as if future Kiribatians have such a right, we might create a situation in which Kiribati will not be flooded.

Let's assume scientists predict that Kiribati will go down in *2065* if worldwide emissions are not curbed from *now* on. As, the life expectancy of Kiribatians is 65 years, we are talking then about the rights of present (instead of future) Kiribatians. I think, these young Kiribatians continue to have a right to a life in their home country *now.* This right might be gone *in 2065* but this does not change the fact that it creates a duty for the world at large *now.* With future generations' future rights, it is just the same: they create duties for us *now.*

16

CAN FUTURE PEOPLE BE SAID TO HAVE LEGAL RIGHTS?

National constitutions

The question of whether there are legal rights of future people is an empirical one. It is correct to speak of their legal rights (and of our legal obligations to them) insofar as these rights are codified in positive law. The task is therefore to browse all national constitutions and the bodies of international law.[1] The increasing acceptance of our responsibility for posterity has resulted in the fact that constitutions and constitutional drafts, especially the ones which were adopted in the last few decades, verbatim refer to generations to come. Among the countries that recently changed their constitutions are France, Germany, Argentina, Brazil, South Africa and many Eastern European countries.[2] All these constitutional clauses of different states, worldwide, can be grouped into three categories: general provisions to protect future people, provisions to protect them in the field of ecology, and provisions to protect them in the field of finances. Many states obviously deem the fields of ecology and finances so prone to intergenerational misconduct that they want to mention them explicitly.[3]

If we analyse all constitutional provisions on rights, duties and responsibilities, we will notice that some speak of basic rights of each citizen while others speak of obligations (or objectives) of the state. These are two diametrically opposed positions. The first position is based on the assumption that conditions for a good life must primarily be maintained for today's generation. If that is achieved, future generations will also benefit from them (*'harmony thesis'*). According to this thesis, there should be an individual basic right to environmental protection to fortify the rights of today's citizens. This attitude is reflected by the constitutions of Argentine, Brazil, Finland, Hungary, Latvia, Portugal, and South Africa, for instance. It does not explicitly mention future generations, but gives every inhabitant of the country the right to a healthy and well-balanced environment. This harmony thesis basically says: 'whatever is good for today's generations is also good for future generations'.[4]

The second position is based on the assumption that here is a conflict of interests between the present and future generations with regard to many environmental aspects, for instance nuclear energy or global warming (*'competition thesis'*). Today's generations can benefit by burdening future generations. It can, for instance, reap the fruits of cheap fossil energy resources while pushing back in time the burdens, the harmful effects of climate change. In this case, a regulation would ideally mention future generations explicitly and underline our responsibility to them. If this competition thesis applies, it would be more appropriate to make the state the guardian of the interests of future people than to introduce a basic right for today's citizens. The German article 20a is based on this approach (it is similar in the Czech Republic, France, Greece, the Netherlands, Lithuania, Spain, Sweden, and Switzerland).

In analysing the protection clauses for future generations in national constitutions, further criteria could be applied, for instance anthropocentrism versus biocentrism. But hardly any constitutional clause is based on a biocentric world view, saying that nature has an intrinsic value, irrespective of its usefulness for man. Finally, it should be pointed out that especially countries with a very rich and glorious history, like Italy or Greece, often mention the preservation of natural and cultural heritage in one breath.

Only three constitutions explicitly grant rights to future generations.[5] Article 11 in the Japanese Constitution of 1946 states that 'these fundamental human rights guaranteed to the people by this Constitution shall be conferred upon the people of this and future generations as eternal and inviolable rights.' Second, Norway's article L 110 b, al 1, as amended in 1992, specifies that 'very person has a right to an environment that is conducive to health and to a natural environment whose productivity and diversity are maintained. Natural resources should be managed on the basis of comprehensive long-term considerations whereby this right will be safeguarded for future generations as well.' Third, article 7 of the Bolivian Constitution, as amended in 2002, states that all citizens have the right 'to enjoy a healthy environment, ecologically well balanced, and appropriate to their well-being, while keeping in mind the rights of future generations.'

Between 2003 and 2009, there has been an attempt to explicitly include a strong reference to the interests of future generations in the German constitution. A think tank named Foundation for the Rights of Future Generations managed to assemble a non-party political group of young members of parliament from all five political parties (CDU/CSU, SPD, FDP and Bündnis 90/The Green Party). With the help of experts from the field of law, the group met during a period of two years to reach a consensus on a text. In the 16th legislative period, 104 members of parliament tabled a motion in the Bundestag for an amendment to the constitution that aimed to anchor intergenerational justice in the constitution. The amendment envisaged supplementing the constitution with a new article:

> Articel 20b (new): 'The state must consider the principle of sustainability and must protect the interests of future generations in its decisions.'

After the first reading of the proposal in the Bundestag (11 October 2007), the draft was sent to several committees for consultation. Once it became clear that the committee on legal affairs – originally leading the consultation process – was not willing to occupy itself with the task, the parliamentary committee for sustainable development took command and organised an expert hearing, which took place on the 15 October 2008. The opinions of the experts with regards to the new article varied. For example, whether Artikel 20b compels policymakers to implement concrete measures was contested. Uncontested, however, was the fact the Artikel 20b had the character of a 'general clause', i.e., not confined in relevance to the ecological and financial sector, but applicable to all sectors in which generational justice is relevant. Hence, at minimum, the clause would apply to political decisions on pensions, health, social care, education and the labour market. With the commencement of the Bundestag's summer break, this initiative – spearheaded by the young members of parliament and the Foundation for the Rights of Future Generations – failed. This is due to the fact that, in Germany, all legislative proposals under consideration are discontinued automatically with the commencement of the summer break.[6]

In Israel, four members of parliament applied for a basic law that should start as follows: 'The objective of this Basic Law is to protect the rights of all people, including those of future generations.'[7]

No matter whether these attempts will be successful in the end or not, the trend indicates that national constitutions will increasingly include rights of future generations.

Of course, such clauses should not speak of *present* legal rights of future people (and none of them does). But all wordings like 'the state should protect the rights of future persons' are unproblematic and avoid the 'present tense/future tense problem' that Beckerman addresses.

'Succeeding' instead of 'future' generations

Another legal innovation is the replacement of the term 'future generations' by 'succeeding generations'. Unlike the term 'future', the term 'succeeding' generations includes not only unborn generations but also present children and adolescents. In many respects, it makes no difference for a theory of an intergenerationally just distribution of resources and life-chances whether a child was born yesterday or will be born tomorrow. In both cases, it has a life to live and should be protected against intergenerational injustice. The generations wielding power today (the middle and the old generation) have the option to use or conserve resources, save or dispose of wealth, secure or neglect institutions – the unborn, as much as the young, do not. The precise age when a young person becomes more powerful is admittedly hard to pin down. A crucial criterion is the voting age.

Imminent future generations and today's infants are on a similar level of powerlessness, thus one could talk about 'succeeding' instead of 'future' generations. If this term 'succeeding generations' were adopted in constitutional provisions or international law, children and adolescents or their parents would have the right to take legal action. The clauses would then have a concrete and judicially guaranteed effect. Then the achievement of the Filipino lawyer Antonio Oposa could be repeated; he successfully sued his government because it did nothing to stop the destruction of the rain forest in the Philippines. Forty-three children (representing succeeding generations) appeared as petitioners. The Federal Constitutional Court of the Philippines upheld the claim of the petitioners on 30 July 1993:

> This case, however, has a special and novel element. Petitioners minors assert that they represent their generation as well as generations yet unborn. We find no difficulty in ruling that they can, for themselves, for others, in their generation and for succeeding generations, file a class suit. Their personality to sue in behalf of succeeding generations can only be cased on the concept of inter-generational responsibility insofar as the right to a balanced and healthful ecology is concerned. [To make the natural resources] equitably accessible to the present as well as to future generations.[8]

Subsequent cases did not follow into the footsteps of this one.[9] But this might change in the future.

International law

It is beyond the scope of this study to provide a comprehensive overview of the provisions to protect future generations in international law.[10] However, a few important legislative landmarks shall be mentioned. Until the Second World War, new fields in international law were introduced according to the principle 'first come, first serve'. Since the

1950s, this principle has been replaced by the doctrine of the common heritage of mankind with regard to a number of territories such as the high seas and the deep sea beds, outer space, the moon, and Antarctica. The doctrine has five principal elements: non-ownership of the heritage, shared management, shared benefits, use exclusively for peaceful purposes, and conservation for mankind.[11] This shift from a Hobbesian to a Kantian element in international law[12] is an important landmark on the way to legal clauses for the protection of the interests of future generations.[13] In 1972 in Stockholm, coming generations were explicitly mentioned at the first UN conference on environmental protection. The Stockholm Declaration (principle 1) reads:

> Man has the fundamental right to freedom, equality and adequate conditions of life, in an environment of a quality that permits a life of dignity and wellbeing, and he bears a solemn responsibility to protect and improve the environment for present and *future generations* [emphasis added].

Ever since then, international declarations have essentially included a reference to the needs of future generations, including those adopted in Rio in 1992 and Johannesburg in 2002. However, we should distinguish such references made in preambles and other non-binding sections of declarations (soft law) from those made in litigable articles of international agreements (hard law). The *UN Convention on Biological Diversity* (1992) is an example for the latter:

> Art. 2.: 'Sustainable Use' means the use of components of biological diversity in a way and at a rate that does not lead to the long-term decline of biological diversity, thereby maintaining its potential to meet *the needs*

and aspirations of present and future generations [emphasis added].

Recently, the principle of 'intergenerational justice' has also been mentioned for the first time, perhaps in order to avoid the question of whether future generations have rights. The *Berlin Commitment for Children*, the closing document of the Conference on Children in Europe and Central Asia preparing for the United Nations General Assembly Special Session on Children (Mai 2001), the term 'intergenerational justice' appears in a document on international law.[14]

On 12 November 1997, the UNESCO adopted a *Declaration on the Responsibilities of the Present Generations Towards Future Generations* at its 29th meeting.[15] It deals primarily with environmental issues (art. 4 and 5). But the protection of the human genome (art. 6), the preservation of peace (art. 9), and education (art. 10) can also be derived from it as objectives of a policy based on generational justice. The declaration was triggered by an initiative of the French marine biologist and ecologist Jacques-Yves Cousteau, who had collected several million signatures for a *Bill of Rights for Future Generations.*[16]

Group rights

It should not go unnoticed that the legal clauses mentioned above speak of 'future generations', thereby referring to groups of people instead of individuals. So, if 'rights' of future generations are mentioned, are these 'group rights'? Let us take a closer look at the concept of group rights.

Group rights are legal rights that all members of a group have in certain countries, solely by virtue of belonging to that group. Affirmative action programmes which grant more 'rights' to members of a particular gender, race, or ethnicity (for instance in the United States) are a topical example. If candidates with the same qualification apply for a job, members of certain groups must be given preference. It is important to note that this concept of group rights is not suitable for the rights of future generations. These kinds of 'group rights' resemble prerogatives more than universal human rights.

In a second view on group rights of future generations, it suffices to think of the rights of future generations as aggregated individual rights. It is not necessary to extend the concept of rights beyond its paradigm application.[17] The concept of rights is usually based on the needs and resulting interests of individuals. Therefore, the rights of a future generation are the same as the rights of the individual members of that generation. There is no difference between the rights of 'all those who live in the year 2300' and the rights of 'the generation of the year 2300', if we define people alive then as a generation.[18] In this view, the rights of groups are the aggregation of the rights of their members.[19]

A third articulation of group rights is grounded in the claim that groups possess interests – and therefore rights. Page explains: 'Many people believe, for example, that the destruction of entire communities or cultures is bad over and above the fact that this is often accompanied by the deaths (or reductions in well-being) of their individual members. On the other hand, people are disposed to view a natural, or human originating, disaster as being more regrettable if it involves the destruction of a whole community than if it involves an identical amount of human misery though dispersed amongst strangers in different communities.'[20]

One of the leading representatives of a concept of rights of future generations in this sense of group rights is Edith Brown-Weiss. The United Nations commissioned her to find out whether international law must be adapted to the global ecological challenges. Her conclusion: 'The thesis of his study is that each generation receives a natural and cultural legacy in trust from previous generations and holds it in trust for future generations. This relationship imposes upon each generation certain planetary obligations to conserve the natural and cultural resource base for future generations and also gives each generation certain planetary rights as beneficiaries of the trust to benefit from the legacy of their ancestors. [. . .] For these obligations and rights to be enforceable, they must become part of international law, and of national and subnational legal systems.'[21] Brown-Weiss proposes a new kind of rights, the 'planetary' or 'intergenerational' rights. 'They are the rights which each generation has to receive the planet in no worse condition than that of the previous generation, to inherit comparable diversity in the natural and cultural resource bases, and to have equitable access to the use and benefits of the legacy.'[22] And then: 'The planetary rights proposed here for future generations are not rights possessed by individuals. Rather they are generational rights, which can only usefully be conceived at a group level.'[23] To explain the nature of this type of rights, Brown-Weiss continues: 'Members of the present generation also possess planetary rights, which are rights derived from membership in the present generation to enjoy the natural resources of earth and our cultural heritage. They derive from intergenerational rights, but are enforced on an intragenerational basis. These rights are associated with corresponding duties, which members of the present generation have towards other members of the same generation. At this stage, they could be viewed as individual rights in the sense that there are identifiable interests of individuals that the rights protect. However, the remedies for violations of these rights will often benefit the rest of the generation, not only the individual, and in this sense they may be said to retain their character as group rights.'[24]

Brown-Weiss explains that she wants to evade the standard objections against rights of future generations that way: 'It has been argued that future generations cannot have rights, because rights exist only when there are identifiable interests, which can happen only if we identify the individuals who have interests to protect. Since we cannot know who the individuals in the future will be, it is not possible for the future generations to have rights. [...] But intergenerational rights are not in the first instance rights possessed by individuals. They are, instead, generational rights, which are held in relation to other generations – past, present and future.'[25]

And as long as there are no inherent contradictions, the ascription of group rights (either type) are also a matter of convention, not of right or wrong. Ultimately, the opinion of the majority is decisive for the allocation of rights. But, since we see no convincing objections against the concept of rights of future generations in the sense of rights of individual future people, we do not think it necessary to allocate group rights to future generations. With regard to our obligations to future people, we have already dealt with the indeterminacy argument that Brown-Weiss seeks to circumvent by postulating group rights. We saw that the indeterminacy argument is not defensible.

17

SUMMARY: IS 'RIGHTS TALK' A GOOD IDEA?

Whom we declare a rights-bearer with regard to a moral right is a question of empathy, of convention and definition. In summary, no logical or conceptual error is involved in speaking about rights of members of future generations. The assertion that there cannot be legal rights for future generations is obviously wrong. Just look into some constitutions and you will find them. Who is a rights-bearer with regard to a legal right is an empirical question.

Obviously a right to emit and a right to a stable climate cannot both be realised at the same time. A conflict between different human rights claims will often be a perplexing dilemma.[1] For example, country A may rely upon the use of fossil fuels to drive its development. On the other hand, country B – with an agrarian economy – may be experiencing dramatic changes in its ability to produce a crop essential for its economic survival due to changing climatic conditions Country A is the cause of this change. In this case, human rights discourse may justify both countries simultaneously. On the one hand, country B has a right to a clean and safe environment. On the other, country A has a right to self-determination and to choose its own development path.[2]

The focus on individual rights gives no guidance as to how such conflicts would be resolved as universal human rights are not tradeable against each other. Rather, such conflicts can only be resolved by a utilitarian approach or variations on liberal principles such as egalitarism, prioritarianism or sufficientism. Human rights alone cannot adequately inform climate policy. One should not ask too much of the human rights approach with regard to climate change.[3] We would not go as far as Mulgan who states: 'The language of rights is problematic, although not impossible with regard to future generations.'[4] But we agree that rights talk is a sometimes tricky way of discussing intergenerational justice and climate change. There are quite convincing rights-based conceptions of justice[5] but they are not necessarily the most appropriate ones. It could be argued that 'rights talk' is a costly distraction, importing further unclarities into the ethical debate and into political UN climate negotiations. Not enough effort has been made to frame theories of intergenerational justice and climate change in other terms, for instance needs-based language.

18

OUTLOOK

Upon printing of this book, the Fifth Assessment Report (AR5) that will be published in 2014 is on the doorstep. Compared with previous reports, the AR5 will put greater emphasis on assessing the socio-economic aspects of climate change and implications for sustainable development, risk management and the framing of a response through both adaptation and mitigation. All the evidence suggests that even the worst-case scenario predictions from AR4 will be exceeded. Climate change drives on faster and faster. After much contentious debate, the international community has finally at least settled on the 2-degree-goal. However scientists are now saying that self-imposed measures taken up to now to protect the climate will in all likelihood lead to a 3.4 to 4 degree rise in average global temperatures. A World bank report called 'Turn down the heat. Why a 4 degree warmer world should be avoided' depicts the severe consequences for the world as we know it if climate change goes on unabated until the end of the century. 'The 4°C scenarios are devastating: the inundation of coastal cities; increasing risks for food production potentially leading to higher malnutrition rates; many dry regions becoming dryer, wet regions wetter; unprecedented heat waves in many regions, especially in the tropics; substantially exacerbated water scarcity in many regions; increased frequency of high-intensity tropical cyclones; and irreversible loss of biodiversity, including coral reef systems.'[1] The president of the world bank starts this report with the words: 'It is my hope that this report shocks us into action.'

The house is burning and its inhabitants are still quarrelling who will call the firefighters and pay their bill. The need for a sound climate ethics is greater than ever.

NOTES

Chapter 1. Introduction: Why an Imminent Threat Stays on the Back Burner

1. Gore 2009, 27.
2. In academic literature, the terms 'global warming' and 'climate change' or 'global climate change' are clearly differentiated. 'Climate change' and 'global climate change' encompass not only higher temperatures but also changes in precipitation patterns and other weather averages. The popular media typically uses these terms synonymously. It is important to know that warming is not the only aspect of climate that is changing. For the sake of text flow, however, these terms will be used interchangeably throughout the book. For more on the terminology, see Conway 2008.
3. Gardiner 2006.
4. Chapter 6–12 are based on Tremmel (2013).
5. World Nuclear Association 2010.
6. Cf. World Nuclear Association 2010, IEA 2010
7. BBC News 2011.
8. It is interesting to note here that these 'guesses' turned out to be astoundingly accurate. By 1895 scientists had predicted what amounts of certain greenhouse gases would result in which climatic phenomena (increase or decrease in atmospheric CO_2 might trigger corresponding glacial advances or retreats). When compared with the findings of the IPCC in 1990, these one-hundred-year-old predictions were very close to the actual results. The 1895 hypotheses

can be seen as a culmination of sorts of sporadic experiments testing if and how people could 'make' the Earth warmer: 'The ability to generate an artificial warming of the Earth's surface was demonstrated in simple greenhouse experiments (. . .) in the 1760s (. . .)' (Treut/Somerville/Cubasch/et al. 2007, 103) In the 1820s scientists realized that air could trap heat as well; in the late 1850s and early 1860s CO_2 was identified as having the potential to alter the climate. Aside from these impressive examples of early research in the natural sciences, climate change remained at best a fringe field of study until the mid- to late twentieth century. For further information on the history of the science of climate, see Treut/Somerville/Cubasch/et al. 2007.

9. Treut/Somerville/Cubasch/et al. 2007, 98.
10. UNDP 2011, 32.
11. Carr/Airlie 2011.
12. Baumert/Herzog/Pershing 2005, 19.
13. Cf. Hamdi-Cherif/Guivarch/Quirion 2011, 733.
14. Cf. Hamdi-Cherif/Guivarch/Quirion 2011, 733.
15. Gore 2009, 27–28.

Chapter 2. The Science of Climate Change

1. Cf. Treut/Somerville/Cubasch/et al. 2007, 96.
2. Cf. Treut/Somerville/Cubasch/et al. 2007, 96.
3. Forster/Ramaswamy/Artaxo/et al. 2007, 135.
4. Cf. Forster/Ramaswamy/Artaxo/et al. 2007, 136.
5. Ramanathan/Carmichael 2008, 222.
6. Ramanathan/Carmichael 2008, 222.
7. Cf. Forster/Ramaswamy/Artaxo/et al. 2007, 163.
8. Ramanathan/Carmichael 2008, 222.
9. Prinn/Reilly/Sarofim/et al. 2007, 94.
10. Prinn/Reilly/Sarofim/et al. 2007, 94.
11. Treut/Somerville/Cubasch/et al. 2007, 108.
12. Kazmin 2009.
13. IPCC 2007, 2.
14. Tans 2012 (see references section for webpage).

15. http://www.sciencebuzz.org/blog/weight-world. Accessed 13 July 2012. To give you a comparison, Mount Everest, the world's tallest mountain, is 8.9 kilometres (5.5 miles) above the surface of the Earth.
16. Forster/Ramaswamy/Artaxo/et al. 2007, 212.
17. Cf. Forster/Ramaswamy/Artaxo/et al. 2007.
18. Cf. Forster/Ramaswamy/Artaxo/et al. 2007.
19. Forster/Ramaswamy/Artaxo/et al. 2007, 212.
20. Forster/Ramaswamy/Artaxo/et al. 2007, 136.
20. Cf. Forster/Ramaswamy/Artaxo/et al.
21. Cf. Forster/Ramaswamy/Artaxo/et al. 2007.
22. Forster/Ramaswamy/Artaxo/et al. 2007, 135.
23. Forster/Ramaswamy/Artaxo/et al. 2007, 141; Tans 2012.
24. Forster/Ramaswamy/Artaxo/et al. 2007, 140.
25. Denman/Brasseur/Chidthaisong/et al. 2007, 539.
26. Forster/Ramaswamy/Artaxo/et al. 2007, 141.
27. Forster/Ramaswamy/Artaxo/et al. 2007, 135.
28. There are long lists of other greenhouse gases, but these typically fall into some larger category, such as halocarbons, chlorofluorocarbons or the like and their contribution to climate change is dwindling as emissions fall due to environmental regulations (cf. Forster/Ramaswamy/ Artaxo/et al. 2007).
29. Denman/Brasseur/Chidthaisong/et al. 2007, 513.
30. Forster/Ramaswamy/Artaxo/et al. 2007, 212.
31. Forster/Ramaswamy/Artaxo/et al. 2007, 141.
32. Denman/Brasseur/Chidthaisong/et al. 2007, 513.
33. Cf. Prinn/Reilly/Sarofim/et al. 2007, 97.
34. Cf. Prinn/Reilly/Sarofim/et al. 2007, 97.
35. Lenton/Held/Kriegler/et al. 2008, 1786.
36. One big problem in establishing emissions limits and climate targets is the level of uncertainty involved in predicting climate change. Based on past measurements, reactions in controlled experiments and complicated models developed over decades, scientists can with varying degrees of certainty predict how the climate will react to changes in the atmospheric consistency. One should be aware, though, that these targets and limits are to a certain extent well-

researched guesswork. Actual changes in climate could be more or less extreme than scientists predict. Based on the IPCC Second Assessment Report, the European Union officially acknowledged 2°C as the highest acceptable rise in global mean temperature due to the anthropogenic climate change in article 6 of the Community Strategy on Climate Change – Council Conclusions section of its 1939th Council Meeting on the Environment in 1996 (EU 1996). The United States and other participants in the UNFCCC 2009 15th Conference of the Parties in Copenhagen formally pledged their agreement that the 2° limit is crucial to preventing 'dangerous anthropogenic interference with the climate system' (UNFCCC 2009, 1).

37. UNFCCC 2012.
38. See, e.g., Gesang 2011, 41; Hadley Centre 2005, 6–7; Elzen, den/Meinshausen 2005, 2. Even at this level, limiting global temperature rise to 2 degrees is not certain. Depending on the source, researchers report that by limiting CO_2 and CO_2^e in the atmosphere to 400 ppm and 450 ppm respectively, we have between a 50 and 75 per cent chance of preventing a higher rise in global mean temperatures.
39. See Gesang 2011, 41–42 (also for further references).
40. IPCC 2007, 5.
41. UNDP 2011, 32.
42. IPCC 2000, 3.
43. A2, A1B and B1 are examples of the possible futures based on the IPCC's development of future scenario 'families' in its Special Report on Emissions Scenarios. For more information on these and other possible future scenarios, see: IPCC (2000).
44. Treut/Somerville/Cubasch/et al. 2007, 111.
45. Treut/Somerville/Cubasch/et al. 2007, 111.
46. UNDP 2011, 35.
47. Treut/Somerville/Cubasch/et al. 2007, 111.
48. Treut/Somerville/Cubasch/et al. 2007, 97.
49. Treut/Somerville/Cubasch/et al. 2007, 105.
50. Treut/Somerville/Cubasch/et al. 2007, 105.
51. Treut/Somerville/Cubasch/et al. 2007, 105.
52. UNDP 2011, 35.

53. Treut/Somerville/Cubasch/et al. 2007, 119.
54. Treut/Somerville/Cubasch/et al. 2007, 122.
55. Treut/Somerville/Cubasch/et al. 2007, 122.

Chapter 3. The Culprits of Climate Change

1. Baumert/Herzog/Pershing 2005, 22.
2. IEA 2009, 41.
3. IEA 2009, 168.
4. Sims/Schock/Adegbululgbe/et al. 2007, 261
5. IEA 2009, 3.
6. IEA 2009, 3.
7. de Sherbinin/Carr/Cassels/Jiang 2007, 15.
8. Denman/Brasseur/Chidthaisong/et al. 2007, 512.
9. See http://www.nationmaster.com/red/pie/env_pol_car_dio_fro_fos_fue_200-carbon-dioxide-fossil-fuels-2000, accessed 19 July 2012.
10. Denman/Brasseur/Chidthaisong/et al. 2007, 512.
11. UNDP 2011, 38.
12. de Sherbinin/Carr/Cassels/Jiang 2007, 6.
13. de Sherbinin/Carr/Cassels/Jiang 2007, 7.
14. de Sherbinin/Carr/Cassels/Jiang 2007, 7
15. UNDP 2011, 38.
16. Forster/Ramaswamy/Artaxo/et al. 2007, 182.
17. Smith/Martino/Cai/et al. 2007, 499.
18. UNDP 2011, 55.
19. UNDP 2011, 55–56.
20. Baumert/Herzog/Pershing 2005, 85.
21. Smith/Martino/Cai/et al. 2007, 501.
22. Smith/Martino/Cai/et al. 2007, 501, 503.
23. Smith/Martino/Cai/et al. 2007, 503.
24. Thus, as population grows, so do rates of deforestation.
25. Data taken from: UN World Population Prospects, the 2010 Revision, http://esa.un.org/unpd/wpp/unpp/p2k0data.asp and the US Census Bureau Population lock, http://www.census.gov/main/www/popclock.html, projected numbers reflect the medium variant of the UN Population Division's projections.

26. With the exception of Europe, where the population is already shrinking. Japan's population is shrinking as well, even as that of other Asian countries continues to grow. Cf. UN Population Division 2004.
27. UNDP 2011, 32.
28. UNDP 2011, 32.
29. Cf., e.g., Satterthwaite 2009; de Sherbinin/Carr/Cassels/Jiang 2007.
30. UNDP 2011, 25.
31. Smith/Martino/Cai/et al. 2007, 500; FAO 2006.
32. Smith/Martino/Cai/et al. 2007, 502.
33. Rosegrant/Paisner/Meijer 2001.
34. Smith/Martino/Cai/et al. 2007.
35. de Sherbinin/Carr/Cassels/Jiang 2007, 13.
36. de Sherbinin/Carr/Cassels/Jiang 2007, 14.
37. UNDP 2011, 40.
38. O'Neill/Dalton/Fuchs/et al. 2010, 17524.
39. O'Neill/Dalton/Fuchs/et al. 2010, 17524.
40. de Sherbinin/Carr/Cassels/Jiang 2007, 13.
41. de Sherbinin/Carr/Cassels/Jiang 2007, 13.
42. de Sherbinin/Carr/Cassels/Jiang 2007, 13.

Chapter 5. Addressing Climate Change: Options and Obstacles

1. See, e.g., Muller/Mendelsohn/Nordhaus 2009; Costanza/d'Arge/Groot/et al. 1997.
2. Muller/Mendelsohn/Nordhaus 2009, 1.
3. Muller/Mendelsohn/Nordhaus 2009, 23.
4. Costanza/d'Arge/Groot/et al. 1997, 283.
5. http://icecap.us/, accessed 18 July 2012.
6. Rahmstorf/Schellnhuber 2007, 31.
7. Denman/Brasseur/Chidthaisong/et al. 2007, 512.
8. http://icecap.us/index.php/go/faqs-and-myths#1, accessed 18 July 2012.
9. D'Aleo 2009.
10. Rahmstorf/Schellnhuber 2007, 36 et seq.
11. See, e.g., Gore 2009; Oreskes/Conway 2010.

12. Associated Press. 'Scientists unsure of absorption.' Accessed from the Global Climate Coalition website on 18 November 2011 (see website address below). All articles the Global Climate Coalition used to question the science of climate change can still be found online in their archives at http://web.archive.org/web/20030408231206/http://globalclimate.org/index.htm, though the organisation itself has since deactivated.
13. ExxonMobil 2005, 22.
14. Adam 2006.
15. Macalister 2006.
16. ExxonMobil 2010.
17. Royal Society 2006.
18. ExxonMobil 2007, 39.
19. The Australian 2010. See also, e.g., Greenpeace 2010; Adam 2009.
20. Kaufman 2011.
21. Mouawad 2009.
22. Kaufman 2011.

Part II. Climate Ethics

1. http://www.guardian.co.uk/environment/2009/dec/19/copenhagen-reaction-delegates-speak. Visited 4.3.10.
2. This includes, for example, the explication of terms, the clarification of arguments, the dialectics, or the conception of thought experiments and analogies that give institutions and logic equal consideration. The method employed in the following chapters is that of thought experiments.
3. Gardiner 2010, 3.
4. Gardiner 2006.

Chapter 6. Distribution of What?

1. Barry 2005, 266–268; Page 2006, 61.
2. Meyer/Roser 2010, 232, respond to this question: 'So, what the shorthand of distributing emissions' ultimately amounts to is distributing (by distribution rights) the benefits of engaging in emission-generating activities. Or, very roughly but more

intuitively, we could say that by distributing emission rights we are distributing economic progress – "very roughly" because, first, economic progress of course does not capture everything that is beneficial about emission-generating activities (sometimes economic progress is not even itself something beneficial), and, second, because there is no one-to-one relationship between emissions and economic progress as some draw much more economic output from the same amount of emissions than others.' It is surprising, then, that Meyer and Roser subsequently equate greenhouse gases with chances for development several times throughout their article, although they themselves cited several reasons for why this is misleading.

3. See Tremmel 2009, 65–146 about indicators for human well-being and quality of living.
4. Miller 2009, 125, 145–151.
5. Miller 2009, 145.
6. For instance Diener/Seligman 2004, 4.
7. And to a lesser extent also the oceans and the biosphere.
8. Scientists use complex computer models to convert CO_2e-emissions into CO_2e-concentrations in the atmosphere and then predict its effects on temperature, sea level and weather. Although these calculations are much more precise than ten years ago, they are still fraught with uncertainties. Estimates of how much greenhouse gas the atmosphere can tolerate, that is, how high the cumulative surge caused by people can be, vary significantly (cf. Hansen et al. 2008; Meinshausen et al. 2006; Solomon et al. 2009; Oleson et al. 2009; WBGU 2009). For the time span until 2050, 560 Gt CO_2e is a plausible mean value. The physical correlations are important but not decisive for the ethical debate. The ethical conclusions derived here apply regardless of whether the environmentally sustainable mass of accumulated CO_2e-emissions is 200 or 800 Gt CO_2e.

   Another point: In the political and ethical debates, one often hears and reads 2 tons of carbon dioxide, not carbon dioxide *equivalents*. For the ethical chain of arguments in this essay, however, this differentiation is not important.

Chapter 7. Intergenerational Justice

1. The exact size of the safe emissions budget can and should be redefined by natural scientists whenever new findings arise.
2. See Page 2006; Partridge 2008.
3. There is broad consensus that the greenhouse effect will incur or has incurred human costs, especially in southern countries. For more details, see Chapter 4 in Part I of this book. The financial costs are estimated by the former chief economist of the World Bank, Nicholas Stern, to be 5 to 20 per cent of global GDP over the next decades (Stern 2007), see also Dasgupta (2006) and Nordhaus (2007).
4. Ludwig 2006, 176.
5. On behalf of the commonalty, see Shue 1999a; Lienkamp 2009, 330–337, and Meyer 2009, 93–96. For an opposing view, see Hillerbrand/Ghil 2008.
6. Kant 2006, 6.
7. Rawls 1971, 319–335.
8. Caney 2010, 220.
9. Shue 2010, 150.
10. Lomborg 2001, 323.
11. This is justified in detail in Tremmel 2009; 2011; 2012.
12. Tremmel 2009, 170.
13. Caney 2010b, 220.
14. For a summary of the arguments, see Tremmel 2009, 71 (with further references).
15. However, Bernward Gesang (2011) points out that there are quasi-discounting factors with regard to climate change, for example the possibility that the effects of a global warming are less severe than is currently feared, or the prospects of climate engineering. According to Gesang's (and our) assessment, even if one estimates these factors highly, it does not change anything on a scale of utilisation that, if we allow an abrupt climate change, we unfairly put our well-being before that of coming generations.

Chapter 8. Pure Distributive Justice

1. 560 Gt divided by the number of people (7 billions) and years (40 years).
2. An issue that is only indirectly related to the question of distribution is the advancement of renewable types of energy. This is undoubtedly necessary and would also be a result of a redistribution of emissions' entitlements.
3. Proponents are, for example: Page 2006, 177–179; Rahmstorf/Schellnhuber 2007, 118; Ott 2003, 196–197; Singer 2002, 39–40; Athanasiou/Baer 2002, 47; Baer 2002, 401; Meyer 2001; Jamieson 2001; Paterson 1996; Shue 1993. The concept of equal distribution per person has been advocated in international politics in the last few decades most notably by developing countries. Of late, even the German chancellor Angela Merkel has committed to this policy; however, she does not want to see it enacted until 2050.
4. Page 2006, 178.
5. I credit Ernest Partridge for this hint. Although I don't share his intuition, it is an important argument.
6. This is the wording in Rawls's book published in 2001, *Justice as Fairness: A Restatement* (Rawls 2001, 42f.).
7. Heinrich-Böll-Foundation 2008.
8. Baer/Athanasiou/Kartha/Kemp-Benedict 2010; Santarius 2009; criticisms of the proposal: Kraus/Ott 2009.
9. Cf. Posner/Weisbach 2010. We differ from their approach significantly, nevertheless, as they reject the role of justice at all in any climate agreement.

Chapter 9. International Justice

1. In the unlikely event that the opposite is proven, we would favour the difference principle over emissions/certificate egalitarianism.
2. For timeliness of data, this figure is a calculation of CO_2 emissions only, i.e., without other greenhouse gases such as methane (CH_4), dinitrogen monoxide (laughing gas, N_2O), chlorofluorocarbons

(CFCs) and sulphur hexafluoride (SF_6). CO_2 emission sources include emissions from energy industry, transport, fuel combustion in industry, services, households, etc, and industrial processes, such as the production of cement. Changes in how land is used can also result in the emission of CO_2, or in the removal of CO_2 from the atmosphere. However, as there is not yet an agreed method for estimating this, it is not included in Figure 9.2.

3. Retrieved 20:13, 30 October 2009 from http://maps.grida.no/go/graphic/total-co2-emissions-from-fossil-fuel-burning-cement-production-and-gas-flaring. Country size is proportionate to national carbon dioxide emissions in 2004. Data for Norway is inaccurate.
4. For instance, a one-hour flight (economy class) amounts to about 0.1 tons of CO_2/person.
5. Rogner/Zhou/Bradley 2007.
6. The political scientist Bernd Guggenberger paints the following picture: 'Let us assume that the chain of human life since the beginning of mankind two to three million years ago comprises 100,000 individual members – one person for each generation who, in his/her size and weight, represents the number of people living in his/her respective generation. The last link of the chain, then, which symbolizes the present generation, is not only the largest and heaviest of all the links, no, it is larger and heavier than all hundred thousand members of the entire chain of mankind before him put together (Guggenberger 27/01/1990).' This example, however, is incorrect. The total number of all people ever living is around 82 billion people. Therefore, the presently living population amounts to about 8.5 per cent of all those who have ever lived. But even this numerical example shows that, from a demographic perspective, we live in unusual times.
7. UNFPA 2009, 21.
8. Alternative data sources, e.g. UN Population Division (2004) or Destatis (2009), present slightly different numbers, but the trend does not change. The CO_2-emissions per person per year are held constant to show the ceteribus paribus effect of population changes. In reality, there may be rebound effects.

9. For more on the ethical legitimacy of population policy, see Tremmel 2008.
10. Line a is drawn as a parallel to the x-axis for reasons of simplicity. In reality, the population only eventually grows from 7 to 9 billions between now and 2050, making the morally legitimate per capital emissions an eventually shrinking variable as well.

Chapter 10. Historical Justice

1. Baumert/Herzog/Pershing 2005, 31, write that industrialised countries' share of historical emissions is 76 per cent and that of developing countries 24 per cent.
2. In the context of historical justice, the term 'low-emissions countries' refers to the accumulated historical emissions, not the current absolute or per capita emissions. Countries with high levels of accumulated emissions frequently, but not always, also have high levels of emissions in the present. Of the 25 highest emitters today, 20 can also be found in the list of highest emitters of all times (Baumert/Herzog/Pershing 2005, 31).
3. See Meyer (2004a, 10). This definition, however, is rejected by Schefczyk, who writes: 'What makes misdeeds historical is not the fact that it happened in the past, or that the people involved are dead; rather, it is the circumstance that the perpetrators know that the public prosecutor or other authorities [regarding the pursuance of this injustice] will remain inactive: (. . .)' (Schefczyk 2009).
4. See among others Miller 2008; Howard-Hassmann 2008; Miller/Kumar 2007; Meyer 2005, 2004a; Caney 2005.
5. Gosseries 2004. We changed the example slightly.
6. Locke 1965, part IV 309, 328–333. Locke postulates hat God gave men in their natural state equal rights for the use of the earth and its resources. He concludes that under this condition, every single person can acquire as much land as he/she wants, given (a) he/she uses what she has appropriated, and (b) he/she leaves enough of the same quantity and quality for others. In this case, the appropriation is not morally objectionable, because others are not worse off than they were before.

7. In addition to the counterarguments discussed here, there is another discussed extensively in the literature: the non-identity problem. This argument, as Tremmel argued elsewhere, is not valid with regard to climate change (see Tremmel 2009, 35–45).
8. Thompson (2009); Kumar/Silver (2004); Lyons (2004); Meyer (2004b); Meyer (2005). This assertion is challenged by other theorists who argue that such compensation or restitution would generate further injustice. The further back the misdeed happened, the more superseding events have taken place. Going back to the original state would produce the sum of so much misery for people living today that it would be a wrong itself, cf., e.g. Waldron (1992) and Sher (1992). The debate (on this aspect cf. also Shue 1999b; Caney 2005, 2010b; and Miller 2009) can still be considered undecided.
9. One could opine that slavery was considered a normal and morally legitimate state of affairs in ancient times by many, if not most, people. The convincing rejoinder (see e.g., Schefczyk 2009) is that a historical agent must still be blamed if he could have evaluated (using standard ethical methods) that such a social practice is immoral.
10. In other words: The production of greenhouse gases is seen perhaps by consequentialists, not, however, by deontologists as bad.
11. Miller 2008, 126–133; Meyer/Roser 2010, 230.
12. Salomon et al. 2007, 824.
13. With the exception of a few notorious climate skeptics, the number of whom dwindle every year. There is, however, still no scientific consensus about the exact effects on temperature, sea level, weather and ultimately the human well-being.
14. Various authors assign different dates to this point (between the 1950s and 1995). 1990 is the date most frequently named, backed, for example, by Singer 2002, 34 or Caney 2010b, 208.
15. Singer 2002, 31.
16. Vanderheiden, 2008, xiii.
17. Ludwig 2006, 134.
18. Ludwig 2006, 135.
19. Fault implies intent or negligence, and neither is present, therefore the UN-Climate Convention speaks not in terms of fault or

guilt, but of responsibility ('Common but differentiated responsibility'), cf. UNFCCC 1992. Preamble, article 3, article 4. This maxim is accepted, at least verbally, by industrialised nations.

20. Meyer/Roser 2010, 235.
21. Höffe 2007, 4.
22. Steinvorth 2007, 12.
23. It is currently subject of some debate whether justice has to be 'relational' or can also be understood as 'non-relational' or alternatively 'non-comparative', see e.g., the anthology by Krebs (2000). However, for the philosophic mainstream it is beyond dispute: 'Poverty is an evil but not itself an injustice' (English 1977, 103).
24. If we consider animals as being moral objects, Robinson Crusoe could theoretically have a justice issue with non-human beings, for instance with some apes over the water resources on his small island.
25. See Meyer 2009, 99; Meyer/Roser 2010, 234.
26. For more on the topic of 'solutions of justice at a single point in time' versus 'solutions of justice for an entire lifespan' see: Holtug/Lippert-Rasmussen 2007.

Chapter 11. From the Currency of Greenhouse Gases to Questions of Monetary Distribution

1. Cf. Stern 2007; Nordhaus 2007, 2008.
2. A flood disaster in which ca. 12 million Pakistanis were affected occurred in August of 2010.
3. Gardiner 2010, 14.
4. Caney 2010b, 205. I diverge from him in the assessment of cases in which it should be used. He also wants to apply it to the cases of the use of the absorptive ability of the atmosphere.
5. Vermeer/Rahmstorf 2009.
6. Caney 2006, 473.
7. Caney 2010b, 208.
8. The is-ought dichotomy is a hot topic in ethics (Hume 1740, 195–212). According to the is-ought dichotomy, it is not possible to seamlessly (or without further prescriptive/normative premises) extrapolate a should-state from the status quo of a relationship or situation.

9. Rothe 2009, 102.
10. Paterson 1996, 133.
11. Rothe 2009, 106. I do not agree with Rothe when he equates 'mitigation' with 'intergenerational equity' and 'adaptation' with 'intra-generational justice'. Intergenerational equity is primarily an anthropocentric concept, and it is unclear whether adaptation or prevention is the better strategy with regard to human welfare. People are very adaptable, animals and plants much less so. Therefore, at least from a biocentric view, the avoidance of greenhouse gases ('mitigation') is preferable. Otherwise, if climate change gains momentum, we would have to expect a massive acceleration of the already high rate of extinction of animals and plants.
12. Proposed Elements of a Protocol to the UNFCCC. Presented by Brazil in Response to the Berlin Mandate (submitted May 1997), FCCC/AGBM/1997/MISC.1/ADD.3. See also La Rovere/de Macedo/Baumert (2002).
13. Cf. also the Lumumba Di-Aping statement that prefaces this section.
14. Der Spiegel, n. 18/2010, p. 131.

Chapter 12. What is Just with Regard to Climate Change?

1. UNFPA 2009, 20.
2. See e.g. Dobson 1996; Goodin 1996; Stein 1998; Lafferty 1998; Kielmansegg 2003; Eckersley 2004; Tremmel 2006; Shearman/Smith 2007; Giddens 2009; Leggewie/Welzer 2009; Thompson 2010.
3. This idea is explicated in detail in Tremmel (2014).

Chapter 13. Rights in the Context of Climate Change: A Contested Terrain

1. A term that appears in the title of an article of Henry Shue (Shue 1993).
2. The IPCC defines 'mitigation' as an 'anthropogenic intervention to reduce the anthropogenic forcing of the climate system; it

includes strategies to reduce greenhouse gas sources and emissions and enhancing greenhouse gas sinks', see Parry et al. (2007), appendix I: glossary, p. 869. We do not discuss here the human rights approach with regard to *adaptation*. For such an approach, see Hall/Weiss (2012); Adger/Paavoli/Huq/Mace (2006).

3. Note that the 'right to a stable climate' is a positive right, but the 'right to greenhouse emissions' is not a negative right. Instead, it is a privilege/power right. Wenar (2011) explains: 'The holder of a negative right is entitled to non-interference, while the holder of a positive right is entitled to provision of some good or service. A right against assault is a classic example of a negative right, while a right to welfare assistance is a prototypical positive right. Since both negative and positive rights are passive rights, some rights are neither negative nor positive. (....) The (privilege-) right to enter a building, and the (power-) right to enter into a binding agreement, are neither negative nor positive.'
4. Tremmel 2009, 143–145, for a summary with further references.
5. For example, by Agarwal/Narain 1991, 5; Shue 1993, 2001, 2005; Vanderheiden 2008, 220; or Lumer 2009.
6. Dworkin 1984, 153.
7. Dworkins justification for 'rights as trumps' is at a same time a critique of utilitarianism: Suppose a community contains a Nazi, for example, whose set of preferences includes the preference that Aryans have more and Jews have less of their preferences fulfilled just because of who they are. A neutral utilitarian [i.e., one that gives the preferences of each individual equal weight; the authors] cannot say that there is no reason in political morality, for rejecting or dishonouring that preference, for not dismissing it as simply wrong, for not striving to fulfil it with all the dedication that officials devote to fulfilling any other sort of preference. Rights as trumps offers a way out of this dilemma. If we determine that every person has the right not to be discriminated against on the base of race or skin colour, then utilitarian calculations are faced with parameters within which to operate (cf. Dworkin 1984, 156).
8. Feinberg 1980, 139. Similar Gosepath 2004, 231.
9. Waldron 1984, 8.

10. Bentham 1795, 334.
11. Ishay 2004, 8.
12. Wenar 2011, 12.
13. Caney 2010a, 165.
14. Previous documents like the British *Bill of Rights* (1689) established rights, esp. of the members of the British Parliament against the crown, but did not declare the universality of rights. (Hunt 2007, 20 et seq.).
15. MacDonald 1984, 21.
16. Burke 1790, § 144.
17. Quoted according to Hunt 2007, 17.
18. De George 1980, 159. On this question, cf. Gewirth 1982.
19. Wenar 2011.
20. For an opposing opinion: Hiskes 2009, 7; cf. also Vlastos 1984; Brandt 1959, 440.
21. Waldron 1984, 1. Cf. also Bentham who favours utilitarian theory over rights talk. He writes: 'The strength of this argument is in proportion to the strength of lungs in those who use it. The principle of utility, with the united powers of Bacon, Locke, Hume, Smith, [and] Paley to develop it, would be nothing against one Danton bawling out natural rights' (Stark 1952, 336).
22. Baier 1980, 182, footnote 1.
23. Hunt 2007, 18.
24. Hunt 2007, 26.
25. Hunt 2007, 26.
26. Current discussions as to what rights (should) exist typically concern socioeconomic rights. Though the International Covenant on Economic, Social and Cultural Rights now has 160 participating parties, the United States continues to reject it.
27. Hunt 2007, 29.
28. Wenar 2011; Hart 1973, 171–179. See also Feinberg 1980, 142 et seq.; Partridge 1990.
29. Another argument is that human-induced climate change has devastating implications on biodiversity. As plants and animals cannot adapt at the same rate as humans to climate change, emitting additional climate gases is directly linked to loss of biodiversity. If ‚rights‘ were attributed to trees and polar bears, they

would use these rights to prevent climate change from happening. But this argument is not discussed in more detail here.

Chapter 14. The 'Present People's Right to a Stable Climate' Approach

1. *Male' Declaration on the Human Dimension of Global Climate Change Male'* (2007).
2. OHCHR Report 2009, 29.
3. OHCHR Report 2009, 23.
4. Inuit Petition 2005.
5. Birnbacher 2009, 128.
6. Lawrence 2013.
7. For more, see Humphreys 2010; Knox 2009, 200–212.
8. Addo 1999.

Chapter 15. The 'Future People's Right to a Stable Climate' Approach

1. I will first discuss the question whether future individuals, not future generations, can have rights. Afterwards, I will deal with the question whether entities such as generations can have rights as groups.
2. Partridge 1980b, 136. Emphasis in the original. Ahrens (1983, 4) also contends that the issue of the rights of future generations is a crucial one. Likewise Hiskes 2009, 7.
3. Birnbacher 1988, 98.
4. Lamont 1946, 83–85.
5. De George 1980, 161; Macklin 1980, 151–152.
6. Beckerman 2006; 2004; 2003; Beckerman/Pasek 2001; Beckerman 1994.
7. Beckerman 2006, 54.
8. Beckerman 2004, 3. Emphasis in the original.
9. Hardly any philosopher explicitly claims that future people will not only have rights in the future, but already have them today, except for Partridge 1990, 54. I disagree with Partridge in this point.

10. Baier 1980, 171. I assume Baier has 'members of generations' in mind.
11. Caney 2010a, 164.
12. Hart distinguishes between 'general rights' and 'special rights', see Hart 1955.
13. Barry 1978, 243.
14. De George 1980, 162.
15. John Locke, (1965), sec. 4, pp. 309 and 328–329, explains that men in the state of nature are moral equals and that God has given to them, in common, the use of the earth and its resources. He claims that, under these conditions, an individual may fairly appropriate land for his own use without belying the equal status of his fellows, provided that he (a) uses rather than wastes what he appropriates and (b) leaves "enough and as good for others". Locke justifies the latter condition on the ground that a person who appropriates a resource, but leaves enough and as good for others, leaves others as well off as they were prior to the appropriation. Hence, they are not injured by his act and have no complaint against him. Given that present and future generations have equal claims to the earth and its resources, Locke's analysis can be extended to apply to the intergenerational allocation of resources.
16. Kavka 1978, 200.
17. Rakowski 1991, 150.
18. Birnbacher 1988, 220.
19. Woodward 1986, 819.
20. Marx 1975, 784.
21. Passmore 1974, 91.
22. Kavka 1978, 188.
23. Partridge 2007, 6.
24. Partridge 1990, 56.
25. Pletcher 1980, 168.
26. Muñiz-Fraticelli 2005, 413.
27. Weston/Bach 2009.
28. Partridge 2007, 6.
29. Cf. UN Population Division 2003, 7.

30. Partridge 1990, 53.
31. Tremmel 2009, 35–46.
32. Thompson (1980) tries to use this objection to explain that we have no obligations to future generations at all. But his arguments are untenable. For a reply, see Ahrens (1983, 2 et seq.; Birnbacher 1988, 152–155; Partridge 2008; Page 2008; Wolf 2008.
33. Feinberg 1973.
34. Partridge 1980a, 2. See also Kavka 1978, 189 et seq.
35. Meyer 2003, 4.
36. Waldron 1984, 2. For this analysis, the terms 'obligation' and 'duty' are used as synonyms.
37. Party A can be a particular person or a state administration.
38. Among others, Narveson 1976, 65; Kavka 1978; Pletcher 1980; Birnbacher 1988; and Waldron 1984, 12, select option 1. Among others, Beckerman 2004; Bandman 1982, 98; and Macklin 1980, 151, cling to option 2. Hart 1984, 80, opts for option 3. Brown-Weiss 1989, 99, chooses option 1 with regard to planetary rights, but option 2 with regard to individual rights.
39. Beckerman 2004, 4. Similar Macklin 1980, 151.
40. Waldron 1984, 10.
41. Bandman 1982, 99. Cf. also Partridge 1990, 42; Waldron 1984, 6.
42. Birnbacher 1988, 100.
43. Cf. Feinberg 1980, 141.
44. Beckerman 2006, 55. I have changed this example by replacing tigers for dodoes and talking about a species becoming extinct today instead of in the past. Among others, De George (1980, 161) and Bandman (1982, 96) also think we do not have the right to any goods if there is no provision to effectively claim this right.
45. The concept of intergenerational compares *average* members of generations (Tremmel 2009, 4). But this does not mean that the concept can only applied on mankind as a whole. It can also be applied to compare generations on a country level, or any other regional level.

Chapter 16. Can Future People be Said to Have Legal Rights?

1. This has been done by Brown-Weiss (1989, 297–328). The task was repeated by Tremmel (2006); Haeberle (2006); Earthjustice (2007, 126–147). For the case of France, see in detail Markus (2012); Bourg (2006).
2. For details, see Tremmel (2006, 192–197). Other countries like Israel, Hungary, or Finland have already set up or are currently discussing new institutions for the protection of future generations instead of including clauses for the protection of future generations in their constitutions. The new institutions are called 'Ombudsman for Future Generations', 'Committee for Future Generations', 'Ecological Council,' 'Future Council,' or 'Third Chamber'; see the articles of Göpel/Arhelger (2010); Ambrusné (2010); Jodoin (2010); Shoham/Lamay (2006); Jávor (2006); Agius (2006).
3. Tremmel 2006, 190.
4. Cf. Beckerman 2006; Wallack 2006.
5. Gosseries 2008, 3.
6. Wanderwitz/Friedrich/Lührmann/Kauch (2008).
7. Shoham/Lamay 2006, 280 et seq.
8. Oposa 2002, 7.
9. Westra 2006, 135.
10. For a good overview, see Weston/Bach 2009.
11. Brown-Weiss 1989, 48. See also Agius 2006, 317–323.
12. Cassese 1996, 31.
13. Westra (2006, 152) criticises that it is allowed to use resources instead of preserving them for future generations. However, if nobody was ever allowed to use non-renewable resources, all generations would lose. Therefore, a sustainable use should be aimed at, for the benefit of all generations. Such a regulation could require each generation to create renewable resources to the same extent as it uses up non-renewable resources, cf. Arrow et al. (1996); Daly (1991); Pearce/Turner (1990).
14. Published as UN document A/AC.256/16. http://www.unicef.org/ceecis/Final_Berlin_Report.pdf. Rev. 8 Aug. 2001.

15. The declaration can be ordered under www.cousteau.org. Rev. 1 September 2011.
16. http://www.cousteau.org/en/cousteau_world/our_programs/future_generations.php.
17. Partridge 1990, 41.
18. See chapter *Comparisons between Generations.*
19. Page 2007, 150; cf. also Jones 2008; Kymlicka 1995.
20. Page 2007, 153.
21. Brown-Weiss 1989, 2.
22. Brown-Weiss 1989, 95. Cf. also Delattre 1972.
23. Brown-Weiss 1989, 96.
24. Brown-Weiss 1989, 96.
25. Brown-Weiss (2002, 5). One of the mentioned critics is Macklin who contents: 'It is common practice to ascribe rights to a class of persons in the legal traditions of some countries and to file class action suits. But the class of persons involved in such a suit is comprised of identifiable individuals' (Macklin 1980, 152).

Chapter 17. Summary: Is 'rights talk' a good idea?

1. Lawrence 2013.
2. Rajamani (2010), points out that there is a conflict between the rights of 500 million poor people in India to gain cheap electricity – required by the right to development – and the right of inhabitants of low-lying Pacific Islanders threatened with inundation to their culture and Islands.
3. Likewise Lawrence 2013; Pedersen 2010.
4. Mulgan 2002, 5.
5. Vlastos 1984.

Chapter 18. Outlook

1. Worldbank 2012, X.

WORKS CITED

Adam, David (1 July 2009): 'ExxonMobil continuing to fund climate sceptic groups, records show'. In: *Guardian*. http://www.guardian.co.uk/environment/2009/jul/01/exxon-mobil-climate-change-sceptics-funding. Accessed 20 December 2011.

——— (20 September 2006): 'Royal Society tells Exxon: stop funding climate change denial'. In: *Guardian*. http://www.guardian.co.uk/environment/2006/sep/20/oilandpetrol.business. Accessed 23 November 2011.

Addo, Michael K. (ed.) (1999): *Human Rights Standards and the Responsibilities of Transnational Corporations*. The Hague: Kluwer Law International.

Adger, W. Neil/Paavoli, Jouni/Huq, Saleemul/Mace, M.J. (eds) (2006): *Fairness in Adaptation to Climate Change*. Cambridge: MIT.

Agarwal, Anil/Narain, Sunita (1991): *Global Warming in an Unequal World: A Case of Environmental Colonialism*. New Delhi: Centre for Science and Environment.

Agius, Emmanuel (2006): 'Intergenerational Justice.' In: Tremmel, Joerg C. (ed.): *Handbook of Intergenerational Justice*. Cheltenham: Edward Elgar Publishing, pp. 317–332.

Ahrens, John (1983): *Preparing for the Future. An Essay on the Rights of Future Generations*. Bowling Green/Ohio: Transaction Publisher.

Ambrusné, Éva T. (2010): 'The Parliamentary Commissioner for Future Generations of Hungary and His Impact'. In: *Intergenerational Justice Review*, Vol. 10 (1), pp. 18–24.

Arrow, Kenneth J./Cline, R./Maler, K.-G./Munasinghe, M./Squitieri, R./ Stiglitz, J. E. (1996): 'Intertemporal Equity, Discounting, and Economic Efficiency'. In: Bruce, J. R./Lee, H./Haites, E. F. (eds): *Climate Change 1995: Economic and Social Dimensions of Climate Change: Contribution of Working Group III to the Second Assessment Report of the Intergovernmental Panel on Climate Change*. Cambridge: Cambridge University Press, pp. 125–144.

Athanasiou, Tom/Baer, Paul (2002): *Dead Heat. Global Justice and Global Warming*. New York: Seven Stories Press.

The Australian (20 July 2010): 'Exxon still aids climate sceptics'. http://www.theaustralian.com.au/news/world/exxon-still-aids-climate-sceptics/story-e6frg6so-1225894256861. Accessed 22 December 2011.

Baer, Paul (2002): 'Equity, Greenhouse Gas Emissions, and Global Common Resources'. In: Schneider, Stephan H./Rosencranz, Armin/Niles, John O. (eds): *Climate Change Policy*. Washington: Island Press, S. pp. 393–408.

Baer, Paul/Athanasiou, Tom/Kartha, Sivan/Kemp-Benedict, Eric (2010): 'Greenhouse Development Rights: A Framework for Climate Protection that is More "Fair" than Equal Per Capita Emission Rights.' In: Gardiner, Stephan M./Caney, Simon/Jamieson, Dale/ Shue, Henry (eds): *Climate Ethics: Essential Readings*. Oxford: OUP, pp. 215–230.

Baier, Annette (1980): 'The Rights of Past and Future Persons'. In: Partridge, Ernest (ed.): *Responsibilities to Future Generations. Environmental Ethics*. Buffalo, N.Y.: Prometheus Books, pp. 171–186.

Bandman, Bertram (1982): 'Do Future Generations Have the Right to Breathe Clean Air?' In: *Political Theory*, Vol. 10 (1), pp. 95–102.

Barry, Brian (1978): 'Circumstances of Justice and Future Generations'. In: Sikora, Richard/Barry, Brian (eds): *Obligations to Future Generations*. Philadelphia: Temple University Press, pp. 205–248.

——— (2005): *Why Social Justice Matters?* Malden, Mass.: Polity Press.

Baumert, Kevin A./Herzog, Timothy/Pershing, Jonathan (2005): 'Navigating the numbers. Greenhouse gas data and international climate policy.' Washington DC: World Resources Institute. http://pdf.wri.org/navigating_numbers.pdf. Accessed 27 April 2012.

BBC News (30 May 2011): 'Global carbon emissions reach record, says IEA.' http://www.bbc.co.uk/news/science-environment-13595174. Accessed 30 May 2011.

Beckerman, Wilfred (2006): 'The Impossibility of a Theory of Intergenerational Justice.' In: Tremmel, Jörg C. (ed.): *Handbook of Intergenerational Justice.* Cheltenham: Edward Elgar Publishing, pp. 53–71.

——— (2004): 'Intergenerational Justice.' In: *Intergenerational Justice Review*, Jg. 4 (2). S. 1–5.

——— (2003): *A Poverty of Reason. Sustainable Development and Economic Growth.* Oakland, CA: The Independent Institute.

——— (1999): 'Sustainability and Intergenerational Equality.' In: Dobson, Andrew (ed.): *Fairness and Futurity: Essays on Environmental Sustainability and Social Justice.* Oxford: Oxford University Press, pp. 71–92.

——— (1994): 'Sustainable Development – Is It a Useful Concept?' In: *Environmental Values*, Jg. 3 (3), S. 101–209.

Beckerman, Wilfred/Pasek, Joanna (2001): *Justice, Posterity, and the Environment.* Oxford: Oxford University Press.

Bentham, Jeremy (1795): 'Supply without Burthen or Escheat vice Taxation.' In: Stark, W. (1952): *Jeremy Bentham's Economic Writings.* Bd. 1. New York: Burt Franklin, pp. 282–367.

Birnbacher, Dieter (2009): 'What does it mean to have a right?' In: *Intergenerational Justice Review*, Jg. 9 (3), pp. 128–132.

——— (1988): *Verantwortung für zukünftige Generationen.* Stuttgart: Reclam.

Bourg, Dominique (2006): 'The French Constitutional Charter for the Environment: An Effective Instrument'? In: Tremmel, Jörg C. (ed.): *Handbook of Intergenerational Justice.* Cheltenham: Edward Elgar Publishing, pp. 230–243.

Brandt, Richard B. (1959): *Ethical Theory.* Englewood Cliffs, New York: Prentice-Hall.

Brown-Weiss, Edith (1989): *In Fairness to Future Generations.* Tokio/New York: United Nations University/Transnational Publishers.

Burke, Edmund (1790): *Reflections on the Revolution in France*. London: Penguin Classics.

Caney, Simon (2010a): 'Climate Change, Human Rights, and Moral Thresholds'. In: Gardiner, Stephan M./Caney, Simon/Jamieson, Dale/Shue, Henry (eds): *Climate Ethics: Essential Readings*. Oxford: OUP, pp. 163–177.

——— (2010b): 'Climate Change and the Duties of the Advantaged'. In: *Critical Review of International Social and Political Philosophy*, Vol. 13 (1), S. pp. 203–228.

——— (2006): 'Environmental Degradation, Reparations, and the Moral Significance of History'. In: *Journal of Social Science*, Vol. 37 (3), pp. 464–482.

——— (2005): 'Cosmopolitan Justice, Responsibility and Global Climate Change'. In: *Leiden Journal of International Law*, Vol. 18, pp. 747–775.

Carr, Mathew/Airlie, Catherine (5 December 2011): 'Global Carbon Emissions Rose by Record Volume Last Year'. In: Bloomberg. http://www.bloomberg.com/news/2011–12–05/global-carbon-emissions-rose-record-amount-in-2010-study-shows.html. Accessed 29 April 2012.

Cassese, Antonio (1996): *International Law in a Divided World*. Oxford: Claredon Press.

Conway, Erik (12 May 2008): 'What's in a Name? Global Warming vs. Climate Change'. http://www.nasa.gov/topics/earth/features/climate_by_any_other_name.html. Accessed 21 May 2012.

Costanza, Robert/d'Arge, Ralph/Groot, Rudolf de/et al. (1997): 'The value of the world's ecosystem services and natural capital'. In: *Nature*, no. 387, pp. 253–260.

Daly, Herman (1991): 'Elements of Environmental Macroeconomics'. In: Costanza, Robert (ed.): *Ecological Economics: The Science and Management of Sustainability*. New York: Columbia University Press, pp. 32–46.

Dasgupta, P. (2006): *Comments on the Stern Review's Economics of Climate Change*. Cambridge, UK: Cambridge University Press.

De George, Richard T. (1980): 'The Environment, Rights, and Future Generations'. In: Partridge, Ernest (ed.): *Responsibilities to Future*

Generations. Environmental Ethics. Buffalo, NY: Prometheus Books, pp. 157–166.

Delattre, Edwin (1972): 'Rights, Responsibilites, and Future Persons'. In: *Ethics*, Vol. 82, pp. 254–258.

Denman, K.L./Brasseur, G./Chidthaisong, A./et al. (2007): 'Couplings between Changes in the Climate System and Biogeochemistry'. In: Solomon, S./Qin, D. /Manning, M. (eds): *Climate Change 2007: The Physical Science Basis. Contribution of Working Group I to the Fourth Assessment Report of the Intergovernmental Panel on Climate Change*. Cambridge, UK/New York, NY, USA: Cambridge University Press, pp. 499–588.

Destatis (2009) (ed.): *Bevölkerung Deutschlands bis 2060. 12. Koordinierte Bevölkerungsvorausberechnung*. Wiesbaden: Statistisches Bundesamt.

Diamond, Jared (1997): *Guns, Germs, and Steel: The Fates of Human Societies*. New York/London: W.W. Norton.

Diener, Ed/Seligman, Martin E. P. (2004): 'Beyond Money: Toward an Economy of Well-being'. In: *Psychological Science in the Public Interest*, Vol. 5 (1), pp. 1–31.

Dobson, Andrew (1996): 'Representative Democracy and the Environment'. In: W. Lafferty/J. Meadowcroft (eds): *Democracy and the Environment*. Cheltenham: Edward Elgar, pp. 124–139.

Dworkin, Ronald (1984): Rights as Trumps. In: Waldron, Jeremy (ed.): *Theories of Rights*. Oxford: Oxford University Press, pp. 153–167.

Earthjustice (2007): Environmental rights report 2007. http://earthjustice.org/sites/default/files/library/references/2007-environmental-rights-report.pdf. Accessed 05 March 2012.

Eckersley, Robyn (2004): *The Green State: Rethinking Democracy and Sovereignty*. Cambridge, Mass./London: MIT.

Elzen, M. G. J. den/Meinshausen, M. (2005): 'Meeting the EU 2°C climate target: global and regional emission implications'. MNP-report 728001031 (www.mnp.nl/en), Netherlands Environmental Assessment Agency (MNP), Bilthoven, The Netherlands. Accessed 13 July 2012.

English, Jane (1977): 'Justice between Generations'. In: *Philosophical Studies*, Vol. 31, pp. 91–104.

European Union (EU) (1996): '1939th Council Meeting – Environment .Brussels, 25 and 26 June 1996 .President: Edo Ronchi, Minister for the Environment of the Italian Republic. PRES/96/188 Date: 26 June 1996'. http://europa.eu/rapid/pressReleasesAction.do?reference=PRES/96/188&format=HTML&aged=1&language. Accessed 12 July 2012.

Exxon. Corporate Citizenship Report 2005 (2005): http://www.exxonmobil.com/Corporate/Files/Corporate/ccr05_fullreport.pdf. Accessed 23 November 2011.

Exxon. Corporate Citizenship Report 2007 (2007): http://www.exxonmobil.com/Corporate/files/Corporate/community_ccr_2007.pdf. Accessed 22 December 2011.

Exxon. Corporate Citizenship Report 2010 (2010): http://www.exxonmobil.com/Corporate/Imports/ccr2010/pdf/community_ccr_2010.pdf. Accessed 22 December 2011

FAO (2006): 'Livestock's long shadow'. ftp://ftp.fao.org/docrep/fao/010/a0701e/a0701e.pdf. Accessed 23 April 2011.

Feinberg, Joel (1980): 'The Rights of Animals and Unborn Generations'. In: Partridge, Ernest (ed.): *Responsibilities to Future Generations. Environmental Ethics.* Buffalo, N.Y.: Prometheus Books, pp. 139–150.

Forster, P./Ramaswamy, V./Artaxo, P./et al. (2007): 'Changes in Atmospheric Constituents and in Radiative Forcing'. In: Solomon, S./Qin, D./Manning, M./et al. (eds). *Climate Change 2007: The Physical Science Basis. Contribution of Working Group I to the Fourth Assessment Report of the Intergovernmental Panel on Climate Change.* Cambridge, UK/New York, NY, USA: Cambridge University Press, pp. 129–234.

Gardiner, Stephan M. (2010): 'Ethics and Global Climate Change'. In: Gardiner, Stephan M./Caney, Simon/Jamieson, Dale/Shue, Henry (eds): *Climate Ethics: Essential Readings.* Oxford: OUP, S. 3–35.

——— (2006): 'A Perfect Moral Storm: Climate change, intergenerational ethics and the problem of moral corruption'. In: *Environmental Values*, Vol. 15, pp. 397–413.

Gardiner, Stephan M./Caney, Simon/Jamieson, Dale/Shue, Henry (2010): *Climate Ethics: Essential Readings.* Oxford: OUP.

Gesang, Bernward (2011): *Klimaethik*. Berlin: Suhrkamp Verlag.

Gewirth, Alan (1982): *Human Rights. Essays on Justification and Applications*. Chicago/London: University of Chicago Press.

Giddens, Anthony (2009): *The Politics of Climate Change*. Polity Press. Cambridge.

Global Climate Coalition website. http://web.archive.org/web/20030408231206/http://globalclimate.org/index.htm. Accessed 22 December 2011.

Goodin, Robert (1996): 'Enfranchising the Earth, and its Alternatives'. In: *Political Studies*, Vol. 44, pp. 835–49.

Göpel, Maja/Arhelger, Malte (2010): 'How to Protect Future Generation's Rights in European Governance'. In: *Intergenerational Justice Review*, Vol. 10 (1), pp. 3–10.

Gore, Al (2009): *Our Choice: A Plan to Solve the Climate Crisis*. Pennsylvania: Rodale Press, Inc.

Gosepath, Stefan (2004): *Gleiche Gerechtigkeit. Grundlagen eines liberalen Egalitarismus*. Frankfurt am Main: Suhrkamp.

Gosseries, Axel (2008): 'On future generations' future rights'. *Journal for Political Philosophy*, Vol. 16 (4), pp. 446–474.

——— (2004): Historical Emissions and Free-Riding. In: Meyer, Lukas H. (ed.): *Justice in Time. Responding to Historical Injustice*. Baden-Baden: Nomos, pp. 355–382.

Greenpeace USA (19 July 2010): Exxon continued to fund climate denial in 2009. http://www.greenpeace.org/usa/en/news-and-blogs/campaign-blog/exxon-continued-to-fund-climate-denial-in-200/blog/26100/. Accessed 20 December 2011.

Guggenberger, Bernd (27 January 1990): Die neue Mehrheit der Lebenden. In: Frankfurter Allgemeine Zeitung. 23/1990 (27 January 1990), gravure print edition, p. 1.

Haeberle, Peter (2006): A Constitutional Law for Future Generations – the 'Other' Form of the Social Contract: The Generation Contract. In: Tremmel, Jörg C. (ed.): *Handbook of Intergenerational Justice*. Cheltenham: Edward Elgar Publishing, pp. 215–229.

Hadley Centre, the (2005): International Symposium on the Stabilisation of Greenhouse Gas Concentrations. Report of the International Scientific Steering Committee. http://stabilisation.metoffice.com/Steering_Commitee_Report.pdf, Accessed 13 July 2012.

Hall, Margaux J./Weiss, David C. (2012): Avoiding Adaptation Apartheid: Climate Change Adaptation and Human Rights Law. In: *Yale Journal of International Law*, 37 (2), pp. 309–366.

Hamdi-Cherif, Meriem/Guivarch, Céline/Quirion, Philippe (2011): Sectoral Targets for Developing Countries: Combining 'Common but Differentiated Responsibilities' with 'Meaningful Participation.' In: *Climate Policy*, Vol. 11(1), pp. 731–751.

Hansen, James/Sato, Makiko/Kharecha, Pushker/Beerling, David/Berner, Robert/Masson-Delmotte, Valerie/Pagani, Mark/Raymo, Maureen/Royer, Dana/Zachos, James (2008): Target Atmospheric CO_2. Where Should Humanity Aim? In: *The Open Atmospheric Science Journal*. No. 2, pp. 217–31.

Hart, Herbert L. A. (1984): Are there any Natural Rights? In Waldron, Jeremy (ed.): *Theories of Rights*. Oxford: Oxford University Press, pp. 77–90.

——— (1973): *Bentham on Legal Rights. Oxford Essays in Jurisprudence*. Second Series. Oxford: Oxford University Press.

——— (1955): Are There Any Natural Rights. In: *Philosophical Review*, Vol. 64 (2), pp. 175–191

Heinrich-Böll-Foundation (2008): The right to development in a climate constrained world. The Greenhouse Development Rights Framework. Berlin. http://www.boell.de/publications/publications-5575.html. Accessed 27 April 2012.

Hillerbrand, Rafaela/Ghil, Michael (2008): Anthropogenic Climate Change. Scientific Uncertainties and Moral Dilemmas. In: *Physica* D 237, 2132–2138. http://www.atmos.ucla.edu/tcd/PREPRINTS/RH&MG-Warming_ethics-Physica_D%2708.pdf. Visited 2. Mai 2010.

Hiskes, Richard P. (2009): *The Human Right to a Green Future: Environmental Rights and Intergenerational Justice*. Cambridge: Cambridge University Press.

Höffe, Otfried (2007): Gerechtigkeit zwischen den Generationen. In: *Intergenerational Justice Review/GenerationenGerechtigkeit!* (German edition), Vol. 7(4), pp. 4–6.

Holtug, Nils/Lippert-Rasmussen, Kasper (2007): An Introduction to Contemporary Egalitarianism. In: Holtug, Nils/Lippert-Rasmussen, Kasper (2007): *Egalitarianism. New Essays on the Nature and Value of Equality*. Oxford: OUP. S. 1–38.

Howard-Hassmann, Rhoda E. (2008): *Reparations to Africa*. Philadelphia: University of Pennsylvania Press.

Hume, David (1740): Ein Traktat über die menschliche Natur. Bd. II. Herausgegeben von Reinhardt Brandt. Hamburg: Meiner (1978), pp. 195–212.

Humphreys, Stephen (ed.) (2010): *Human Rights and Climate Change*. Cambridge: Cambridge University Press.

Hunt, Lynn (2007): *Inventing Human Rights*. New York: W.W. Norton & Company.

IEA (2010): Nuclear phase-out. http://www.iea.org/press/pressdetail.asp?PRESS_REL_ID=229. Accessed 10 September 2011.

Inuit Petition – Petition to the Inter-American Commission on Human Rights seeking Relief from Violations Resulting from Global Warming caused by Acts and Emissions of the United States (7 December 2005). www.inuitcircumpolar.com/files/uploads/icc-files/FINALPetitionICC.pdf. Accessed 19 July 2012.

IPCC (2007): Summary for Policymakers. In: Solomon, S./Qin, D./Manning, M./et al. (eds): *Climate Change 2007: The Physical Science Basis. Contribution of Working Group I to the Fourth Assessment Report of the Intergovernmental Panel on Climate Change*. Cambridge, United Kingdom and New York, NY, USA: Cambridge University Press, pp. 1–18.

IPCC (2000): Summary for Policymakers: Special Report on Emissions Scenarios. In: IPCC (Nakicenovic, Nebojsa/Alcamo, Joseph/Davis, Gerald/et al. (Lead Authors)): IPCC Special Report on Emissions Scenarios. A Special Report of Working Group III of the Intergovernmental Panel on Climate Change. Cambridge, United Kingdom and New York, NY USA, pp. 1–20.

IPCC (1990): Climate Change: The IPCC Scientific Assessment (1990). In: Houghton, J.T./Jenkins, G.J./Ephraums, J.J. (eds): *Contribution of Working Group I to IPCC First Assessment Report (FAR)*. Cambridge: Cambridge University Press.

International Energy Agency (IEA) (2009): World Energy Outlook. http://www.iea.org/textbase/nppdf/free/2009/weo2009.pdf. Accessed 12 July 2012.

Ishay, Micheline R. (2004): *The History of Human Rights. From Ancient Times to the Globalization Era*. Berkeley: University of California Press.

Jamieson, Dale (2001): Climate Change and Global Environmental Justice. In: Miller, Clark A./ Edwards, Paul N. (eds): *Changing the Atmosphere: Expert Knowledge and Environmental Governance.* Cambridge/London: MIT Press, pp. 247–286.

Jávor, Benedek (2006): Institutional Protection of Succeeding Generation-Ombudsman for Future Generations in Hungary. In: Tremmel, Jörg C. (ed.): *Handbook of Intergenerational Justice.* Cheltenham: Edward Elgar Publishing, pp. 282–298.

Jodoin, Sébastien (2010): Crimes against Future Generations: Implementing Intergenerational Justice through International Criminal Law. In: *Intergenerational Justice Review*, Vol. 10(1), pp. 11–17.

Jones, Peter (2008): Group Rights. In: *Stanford Encyclopedia of Philosophy.* http://plato.stanford.edu/entries/rights-group/. Accessed 18 July 2012.

Kant, Immanuel (2006): *Toward Perpetual Peace and Other Writings on Politics, Peace, and History.* Edited and with an introduction by Pauline Kleingeld; translated by David L. Colclasure. New Haven, Conn.: Yale University Press. (German original: *Idee zu einer allgemeinen Geschichte in weltbürgerlicher Absicht.* Göttingen/ Hamburg: Verlag Öffentliches Leben GmbH 1949, p.6., German original first published 1785).

Kaufman, Leslie (1 July 2011): Exxon Cut Financing to Climate Skeptics, Group Says. In: *The New York Times.* http://green.blogs.nytimes.com/2011/07/01/exxon-is-financing-fewer-climate-skeptics-group-says/. Accessed 20 December 2011.

Kavka, Gregory S. (1978): The Futurity Problem. In: Sikora, Richard / Barry, Brian (eds.) (1978): *Obligations to Future Generations.* Philadelphia: Temple University Press, pp. 186-203.

Kazmin, Amy (22 September 2009): Rising Sea Levels Hit Bangladesh Livelihoods. In: *Financial Times*, p. 5.

Kielmansegg, Peter Graf (2003): Können Demokratien zukunftsverantwortlich handeln? In: *Merkur*, Vol. 57(7), pp. 583–594.

Knox, John (2009): Climate Change and Human Rights Law. In: *Virginia Journal of International Law*, Vol. 50, 163–218.

Kovats, R.S./Campbell-Lendrun, D./Matthies, F. (2005): Climate Change and Human Health: Estimating Avoidable Deaths and Disease. In: *Risk Analysis*, Vol. 25, pp. 1409–1418.

Kraus, Katrin/Ott, Konrad (2009): Contraction and Convergence. Streitgespräch: Wie fair ist fair genug? Zwei Klimakonzepte im Vergleich. In: *BÖLL.Thema*, no. 2, pp. 33–35.

Krebs, Angelika (ed.) (2000): *Gleichheit oder Gerechtigkeit? Texte der neuen Egalitarismuskritik*. Frankfurt am Main: Suhrkamp.

Kumar, Rahul/Silver, David (2004): 'The Legacy of Injustice. Wronging the Future, Responsibility for the Past'. In: Meyer, Lukas H. (ed.): *Justice in Time. Responding to Historical Injustice*. Baden-Baden: Nomos, pp. 145–159.

Kymlicka, Will (1995): *Multicultural Citizenship*. Oxford: Oxford University Press.

Lafferty, William M. (1998): 'Democracy and Ecological Rationality: New Trials for an Old Ceremony. Aufsatz für CPSA/IPSA Roundtable. Quebec City. 12 February 1998'. http://www.prosus.uio.no/publikasjoner/Andre_pub/1.html. Accessed 30 January 2009.

Lamont, William D. (1946): *The Principles of Moral Judgement*. Oxford: Clarendon Press.

La Rovere, E./de Macedo, L./Baumert, K. (2002): 'The Brazilian Proposal on Relative Responsibility for Global Warming'. In: Baumert, K./Blanchard, O./Llosa, S./Perkaus, J. (eds): *Building on the Kyoto Protocol: Options for Protecting the Climate*. Washington: World Resource Institute Publications.

Lawrence, Peter (2013) *Justice for Future Generations, Climate Change and International Law*. Cheltenham: Edward Elgar Press.

Leggewie, Claus/Welzer, Harald (2009): *Das Ende der Welt, wie wir sie kannten. Klima, Zukunft und die Chancen der Demokratie*. Frankfurt am Main: Fischer.

Lenton, T. M./Held, H./Kriegler, E./ Hall, J. W./Lucht, W./Rahmstorf, S./Schellnhuber, J. (2008): 'Tipping Elements in the Earth's Climate System'. In: *Proceedings in the National Academy of Science*, Vol. 105 (6), pp. 1786-1793.

Lienkamp, Andreas (2009): *Klimawandel und Gerechtigkeit: Eine Ethik der Nachhaltigkeit in christlicher Perspektive.* Paderborn: Ferdinand Schöningh.

Locke, John (1965): *Two Treatises of Government.* Edited by Peter Laslett. Second Treatise. New York: New American Library (first published 1689).

Lomborg, Bjørn (2001): *The Skeptical Environmentalist: Measuring the Real State of the World.* Cambridge: Cambridge University Press.

Long, Graham (2011): 'Disagreement and Responses to Climate Chance'. In: *Environmental Values.* Vol. 20, pp. 503–525.

Ludwig, Karl-Heinz (2006): *Eine kurze Geschichte des Klimas. Von der Entstehung der Erde bis heute.* München: Beck'sche Reihe.

Lumer, Christoph (2009): Klimawandel, 'Generationengerechtigkeit und Entwicklung'. In: *Intergenerational Justice Review,* Vol. 3 (9), pp. 88–95.

Lyons, David (2004): 'Unfinished Business. Racial Junctures in US History and Their Legacy'. In: Meyer, Lukas H. (ed.): *Justice in Time. Responding to Historical Injustice.* Baden-Baden: Nomos, pp. 271–299.

MacDonald, Margaret (1984): 'Natural Rights'. In: Waldron, Jeremy (ed.): *Theories of Rights.* Oxford: Oxford University Press, pp. 21–40.

Macklin, Ruth (1980): 'Can Future Generations Correctly Be Said to Have Rights?' In: Partridge, Ernest (ed.): *Responsibilities to Future Generations. Environmental Ethics.* Buffalo, N.Y.: Prometheus Books, pp. 151–157.

Male Declaration on the Human Dimension of Global Climate Change Male, Republic of Maldives, 14 November 2007. www.ciel.org/Publications/Male_Declaration_Nov07.pdf. Accessed 14 May 2012.

Markus, Jean-Paul (2012): *Quelle responsabilité juridique envers les générations futures?* Paris: Dallos.

Marx, Karl (1975): 'Das Kapital. Kritik der politischen Ökonomie. Vol. 3'. In: Marx, Karl / Engels, Friedrich: Werke. Vol. 25. Berlin: Dietz (first published in 1894).

Meehl, G.A./Stocker, T.F./Collins, W.D./et al. (2007): 'Global Climate Projections'. In: Solomon, S./Qin, D./Manning, M./et al. (eds): *Climate Change 2007: The Physical Science Basis. Contribution of Working Group I to the Fourth Assessment Report of the Intergovernmental Panel on Climate Change.* Cambridge, United Kingdom and New York, NY, USA: Cambridge University Press, pp. 747–846.

Meinshausen, Malte (2005): 'Emission & concentration implications of long-term climate targets'. http://e-collection.library.ethz.ch/eserv/eth:27896/eth-27896–01.pdf. Accessed 14 July 2012.

Meinshausen, Malte/Hare, Bill/Wigley, Tom/Van Vuuren, Detlef/Den Elzen, Michael/Swart, Rob (2006): 'Multi-Gas Emissions Pathways to Meet Climate Targets'. In: *Climatic Change*, Vol. 75 (1), pp. 151–94.

Meyer, Aubrey (2001): *Contraction & Convergence. The Global Solution to Climate Change.* Totnes: Green Books.

Meyer, Lukas H./Roser, Dominic (2010): 'Climate Justice and Historical Emissions'. In: *Critial Review of International Social and Political Philosophy*, Vol. 13 (1), pp. 229–253.

——— (2009): 'Klimawandel und Gerechtigkeit'. In: *Transit* 36 (Winter 08/09), pp. 80–106.

——— (2005): *Historische Gerechtigkeit.* Berlin/New York: de Gruyter.

——— (2004a): 'Einleitung'. In: *Meyer*, Lukas H. (ed.): *Justice in Time. Responding to Historical Injustice.* Baden-Baden: Nomos, pp. 9–52.

——— (2004b): 'Compensating Wrongless Historical Emissions of Greenhouse Gases'. In: *Ethical Perspectives*, no. 11, pp. 22–37.

Miller, David (2009): 'Global Justice and Climate Change: How Should Responsibilities be Distributed?' *Tanner Lectures on Human Values*, 28, S. 117–156.

——— (2008): *National Responsibility and Global Justice.* Oxford: Oxford University Press.

Miller, Jon/Kumar, Rahul (eds) (2007): *Reparations. Interdisciplinary Inquiries.* Oxford: Oxford University Press.

Muller, Nicholas/Mendelsohn, Robert/Nordhaus, William (2009): 'Environmental accounting for pollution: methods with an application to the United States economy'. http://www.nber.org/public_

html/confer/2009/SI2009/EEE/Muller_Nordhaus_Mendelsohn.pdf. Accessed 1 December 2011.

Muñiz-Fraticelli, Victor M. (2005): 'Book Review: Gosseries, Axel: Penser la justice entre les générations: De l'affaire Perruche à la réforme de retraites'. In: *Ethics,* Jg. Vol. 115 (2), S. 412–415.

Narveson, Jan (1976): 'Moral Problems of Population'. In: Bayles, Michael D. (ed.): *Ethics and Population.* Cambridge: Schankman, pp. 59–80.

Nordhaus, William (2008): *A Question of Balance. Weighing the Options on Global Warming Policies.* Yale: Yale University Press.

——— (2007): 'Critical Assumptions in the Stern Review on Climate Change'. *Science*, Vol. 317 (5835), pp. 201–202.

OHCHR report – Report on the Office of the United Nations High Commissioner for Human Rights on the Relationship between Climate Change and Human Rights (2009). Annual Report of the United Nations High Commissioner for Human Rights and Reports of the Office of the High Commissioner and the Secretary-General, Human Rights Council, 15 January 2009, UN Doc. A/HRC/10/61.

Oleson, Kirste/Hartzell, Lauren/Mastrandrea, Michael D. (2009): 'The Baker's Dozen: Key Nations Can and Should Act Together to Prevent Further Dangerous Climate Change'. In: *Intergenerational Justice Review*, 3 (9), pp. 106–112.

O'Neill, Brian C./Dalton, Michael/Fuchs, Regina/et al. (2010): 'Global Demographic Trends and Future Carbon Emissions'. In: *Proceedings of the National Academy of Sciences*, Vol. 107 (41), pp. 17521–17526.

Oposa, Antonio (2002): 'In Defence of Future Generations'. In: *Intergenerational Justice Review* (English edition), Vol. 2 (3), p. 7.

Oreskes, Naomi (17 April 2008): 'You can argue with the facts: the denial of global warming'. Presentation, Stanford University. Available at http://www.aip.org/history/powerpoints/GlobalWarming_Oreskes.ppt

Oreskes, Naomi/Conway, Eric M. (2010): *Merchants of Doubt: How a Handful of Scientists Obscured the Truth on Issues from Tobacco Smoke to Global Warming.* New York: Bloomsbury Press.

Ott, Konrad (2003): 'Ethische Aspekte des Klimawandels'. In: Gottschalk-Mazouz, Niels/Mazouz, Nadia (eds): *Nachhaltigkeit und Globaler Wandel. Integrative Forschung zwischen Normativität und Unsicherheit.* Frankfurt/ New York: Campus, pp. 169–201.

Page, Edward (2008): 'Three problems of intergenerational justice'. In: *Intergenerational Justice Review*, Vol. 8 (1), pp. 9–12.

——— (2006): *Climate Change, Justice and Future Generations.* Cheltenham: Edward Elgar.

Parry, Martin/Canziani, Osvaldo/Palutikof, Jean/van der Linden, Paul/ Hanson, Cair (2007): *Climate Change 2007: Impacts, Adaptation and Vulnerability-Contribution of Working Group II to the Fourth Assessment Report of the IPCC.* Cambridge: Cambridge University Press.

Partridge, Ernest (2008): 'Just Provision for the Future'. In: *Intergenerational Justice Review* (English edition), Vol. 8 (1), pp. 4–8.

——— (2007): 'Should we seek a better future?' www.igc.org/gadfly/ papers/swsabf.htm. Rev. 4 January 2008.

——— (1990): 'On the Rights of Future Generations'. In: Scherer, Donald (ed.): *Upstream-Downstream. Issues in Environmental Ethics.* Philadelphia: Temple University Press, pp. 40–66.

——— (1980a): 'Introduction'. In: Partridge, Ernest (ed.): *Responsibilities to Future Generations.* Buffalo, N.Y.: Prometheus Books, pp. 1–16.

——— (1980b): 'Why Care About the Future?' In: Partridge, Ernest (ed.): *Responsibilities to Future Generations.* Buffalo, N.Y.: Prometheus Books, pp. 203–220.

——— (ed.) (1980c): *Responsibilities to Future Generations.* Buffalo, N.Y.: Prometheus Books.

Passmore, John (1974): *Man's Responsibility for Nature.* London: Duckworth.

Paterson, Matthew (1996): *Global Warming and Global Politics.* London/ New York: Routledge.

Pearce, David/Turner, Kerry R. (1990): *Economics of Natural Resources and the Environment.* London: Harvester Wheatsheaf.

Pedersen, Ole W. (2010): 'Climate Change and Human Rights: Amicable or Arrested Development?' In: *Journal of Human Rights and the Environment*, Vol. 1 (2), pp. 236–258.

Pletcher, Galen (1980): 'The Rights of Future Generations'. In: Partridge, Ernest (ed.): *Responsibilities to Future Generations. Environmental Ethics.* Buffalo, N.Y.: Prometheus Books, pp. 167–170.

Posner, Eric. A./ Weisbach, David (2010): *Climate Change Justice.* Princeton: Princeton University Press.

Prinn, Ronald/Reilly, John M./Sarofim, Marcus/et al. (2007): 'Effects of air pollution control on climate: results from an integrated global system model'. In: Schlesinger, Michael/Kheshgi, Haroon/Smith, Joel/et al. (eds). *Human-induced Climate Change: An Interdisciplinary Assessment.* Cambridge: Cambridge University Press, pp. 93–102.

Rahmstorf, Stefan/Schellnhuber, Hans Joachim (2007): *Der Klimawandel. Diagnose, Prognose, Therapie.* München: Beck.

Rajamani, Lavanya (2010): 'The Increasing Currency and Relevance of Rights-Based Perspectives in the International Negotiations on Climate Change'. In: *Journal of Environmental Law*, Vol. 22(3), pp. 391–429.

Rakowski, Eric (1991): *Equal Justice.* Oxford: Clarendon Press.

Ramanathan, V./Carmichael, G. (2008): 'Global and Regional Climate Changes Due to Black Carbon'. In: *Nature Geoscience*, no. 1, pp. 221–227.

Rawls, John (1971): *A Theory of Justice.* Cambridge, Mass.: Harvard University Press.

Rogner, H.-H./Zhou, D/Bradley, R./et al. (2007): 'Introduction'. In: Metz, B./Davidson, O.R./Bosch, P.R./et al. (eds): *Climate Change 2007: Mitigation. Contribution of Working Group III to the Fourth Assessment Report of the Intergovernmental Panel on Climate Change.* Cambridge, UK/New York, NY, USA: Cambridge University Press.

Rosegrant, M./Paisner, M.S./Meijer S. (2001): Long-Term Prospects for Agriculture and the Resource Base. The World Bank Rural Development Family. Rural Development Strategy Background Paper #1. The World Bank, Washington.

Rothe, Delf (2009): 'Just Adaptation? How the Diffusion of Norms in the Global Climate Regime Affects International'. In: *Intergenerational Justice Review*, Vol. 9 (3), pp. 100–105.

Salomon, Susan/Qin, D./Manning, M./Chen, Z./Marquis, M./Averyt, K. B./Tignor, M./Miller, H. L. (eds) (2007): *Climate Change 2007: The Physical Science Basis. Contribution of Working Group I to the Fourth Assessment Report of the Intergovernmental Panel on Climate Change.* Cambridge.

Santarius, Tilman: 'Contraction and Convergence. Streitgespräch: Wie fair ist fair genug? Zwei Klimakonzepte im Vergleich'. In: *BÖLL. Thema*, no. 2, pp. 35–36.

Satterthwaite, David (2009): 'The Implications of Population Growth and Urbanization for Climate Change'. In: *Environment and Urbanization*, Vol. 21, pp. 545–568.

Schefczyk, Michael (2009): 'Der Unterschied zwischen historischem Unrecht und historischem Übel'. In: *Journal für Generationengerechtigkeit*, Vol. 9 (1), pp. 4–9.

Sher, George (1992): 'Ancient Wrongs and Modern Rights'. In: Laslett, Peter/Fishkin, James S. (eds): *Justice between Age Groups and Generations*. New Haven: Yale University Press, pp. 48–61.

Shearman, David/Smith, Joseph W. (2007): *The Climate Change Challenge and the Failure of Democracy*. Westport: Praeger Publishers

Sherbinin, Alex de/Carr, David/Cassels, Susan/Jiang, Leiwen (2007): 'Population and Environment'. In: *Annu Rev Environ Resour*, no. 32, pp. 345–373.

Shoham, Shlomo/Lamay, Nira (2006): 'Commission for Future Generations in the Knesset: Lessons Learnt'. In: Tremmel, Jörg C. (ed.): *Handbook of Intergenerational Justice*. Cheltenham: Edward Elgar Publishing, pp. 244–281.

Shue, Henry (2010): 'Deadly Delays, Saving Opportunities'. In: Gardiner, Stephan M./Caney, Simon/Jamieson, Dale/Shue, Henry (2010): *Climate Ethics: Essential Readings*. Oxford: OUP, S. 146–162.

——— (2005): Responsibility of Future Generations and the Technological Transition. In: Sinnott-Armstrong, Walter/Howarth, Richard (eds.): *Perspectives on Climate Change: Science, Economics, Politics, Ethics*. Elsevier, pp. 265–284.

——— (2001): Climate. In: Jamieson, Dale (ed.): *A Companion to Environmental Philosophy*. Malden, MA: Blackwell, pp. 449–459.

——— (1999a): Bequeathing Hazards: Security Rights and Property Rights of Future Humans. In: Dore, M/Mount, T (ed.: *Global Environmental Economics: Equity and the Limits to Markets*. Oxford: Blackwell, S. 38–53.

——— (1999b): Global Environment and International Inequality. In: *International Affairs*, Vol. 75, S. 531–545.

——— (1993): Subsistence Emissions and Luxury Emissions. In: *Law & Policy*, Vol. 15(1), S. 39–59.

Singer, Peter (2002) (ed.): *One World: The Ethics of Globalization.* New Haven, CT: Yale University Press.

Smith, P. D./Martino, Z./Cai, D./et al. (2007): Agriculture. In: Metz B./Davidson, O.R./Bosch P.R./et al. (eds.): *Climate Change 2007: Mitigation. Contribution of Working Group III to the Fourth Assessment Report of the Intergovernmental Panel on Climate Change.* Cambridge, United Kingdom/New York, NY, USA: Cambridge University Press.

Solomon, Susan/Plattner, Gian-Kasper/Knutti, Reto/Friedlingstein, Pierre (2009): Irreversible Climate Change Due to Carbon Dioxide Emissions. In: *Proceedings of the National Academy of Sciences*, Vol. 106(6), pp. 1704–1709.

Stark, W. (1952): *Jeremy Bentham's Economic Writings.* Vol. I. London: Allen & Unwin.

Stein, Tine (1998): *Demokratie und Verfassung an den Grenzen des Wachstums. Zur ökologischen Kritik und Reform des demokratischen Verfassungsstaates.* Opladen.

Steinvorth, Ulrich (2007): 'Generationengerechtigkeit'. In: *Intergenerational Justice Review/GenerationenGerechtigkeit!* (German edition), Vol. 7 (4), pp. 12–15.

Stern, Nicolas (2007): *The Economics of Climate Change. The Stern Review.* Cambridge: Cambridge University Press.

Tans, Pieter Dr. (2012): National Oceanic and Atmospheric Administration (NOAA)/Earth Science Research Laboratory (ESRL) data set: Concentrations of CO2 in the Earth's atmosphere (parts per million) derived from in situ air measurements at the Mauna Loa Observatory, Hawaii. www.esrl.noaa.gov/gmd/ccgg/trends/. Accessed 10 July 2012.

Thompson, Dennis F. (2010): 'Representing future generations: political presentism and democratic trusteeship'. In: *Critical Review of International Social and Political Philosophy.* Heft 1 (13. Jg.), pp. 17–37.

Thompson, Janna (2009): *Intergenerational Justice. Rights and Responsibilities in an Intergenerational Polity.* New York: Routledge.

Thompson, Thomas (1980): 'Are We Obligated to Future Others?' In: Partridge, Ernest (ed.): *Responsibilities to Future Generations. Environmental Ethics*. Buffalo, N.Y.: Prometheus Books, 195–202.

UN Population Division (2003): *World Population in 2300*. New York: UN Publications.

Tremmel, Joerg, C. (2014): 'An Extended Separation of Powers Model as the Theoretical Basis for the Representation of Future Generations'. In: Birnbacher, Dieter/Thorseth, May (ed.): *Roads to Sustainability* (forthcoming).

——— (2013): 'Climate Change and Political Philosophy: Who Owes What to Whom?' In: *Environmental Values*. http://www.whpress.co.uk/EV/EVpapers.html. Viewed on 15 January 2013.

——— (2012): *Eine Theorie der Generationengerechtigkeit*. Münster: mentis.

——— (2011): Haben wir die Welt nur von unseren Kindern geborgt? Grundzüge einer Generationenethik. In: Ach, Johann/Bayertz, Kurt/Siep, Ludwig (ed): *Grundkurs Ethik. Band II. Die Bereichsethiken*. Paderborn: mentis, pp. 99–116.

——— (2009): *A Theory of Intergenerational Justice*. London: Earthscan.

——— (2008): An Ethical Assessment of the Legitimacy of Anti-natalistic Birth Policies. In: Tremmel, Joerg C. (ed.): *Demographic Change and Intergenerational Justice. The Implementation of Long-term Thinking in Political Decision-Making*. Berlin/Heidelberg: Springer, pp. 137–159.

——— (2006): Establishment of Intergenerational Justice in National Constitutions. In: Tremmel, Joerg C. (ed.): *Handbook of Intergenerational Justice*. Cheltenham: Edward Elgar Publishing. pp. 187–214.

Treut, H. Le/Somerville, R./Cubasch, U./et al. (2007): Historical Overview of Climate Change. In: Solomon, S./Qin, D./Manning, M./et al. (eds): *Climate Change 2007: The Physical Science Basis. Contribution of Working Group I to the Fourth Assessment Report of the Intergovernmental Panel on Climate Change*. Cambridge, United Kingdom/New York, NY, USA: Cambridge University Press, pp. 93–128.

United Nations Development Programme (UNDP) (2011): Human Development Report 2011. Basingstoke: Palgrave Macmillan. http://hdr.undp.org/en/reports/global/hdr2011/download/. Accessed 5 April 2012.

UNFCCC (2012): Essential Background. Basic Facts and Figures. http://unfccc.int/essential_background/basic_facts_figures/items/6246.php. Accessed 11 July 2012.

——— (2009): Copenhagen Accord. http://www.eenews.net/public/25/13656/features/documents/2009/12/19/document_gw_01.pdf. Accessed 12 July 2012.

——— (1992): Rahmeneinkommen der Vereinten Nationen über Klimaänderungen. http://unfccc.int/resource/docs/convkp/convger.pdf. Zugriff am 19 August 2009.

UNFPA (2009): *State of the World Population 2009. Facing a Changing World: Women, Population and Climate*. New York: UN Publications.

United Nations Population Division (2010): 'World Population Prospects: The 2010 Revision'. http://esa.un.org/unpd/wpp/index.htm. Accessed 04 June 2012.

——— World Population Prospects. The 2008 Revision. http://esa.un.org/unpd/wpp2008/index.htm

——— (2004): 'World Population to 2300'. http://www.un.org/esa/population/publications/longrange2/WorldPop2300final.pdf

United Nations Population Fund/International Institute for Environment and Development (UNFPA/IIED) (eds) (2009): *Population Dynamics and Climate Change*. Edited by José Miguel Guzmán, George Martine, Gordon McGranahan, Daniel Schensul, Cecilia Tacoli. New York: UN Publications.

United States Census Bureau (2012): U.S. and World Population Clocks. http://www.census.gov/main/www/popclock.html. Accessed 04 June 2012.

United States Environmental Protection Agency (USEPA) (2010): Our Nation's Air – Status and Trends Report through 2008. http://www.epa.gov/airtrends/2010/report/fullreport.pdf. Accessed on 01 March 2012.

Vanderheiden, Steve (2008): *Atmospheric Justice. Political Theory of Climate Change*. Oxford: OUP.

Vermeer, Martin/Rahmstorf, Stefan (2009): 'Global Sea Level Linked to Global Temperature'. In: *Proceedings of the National Academy of Science*. http://www.pnas.org/content/early/2009/12/04/0907765106.full.pdf. Accessed on 28 December 2010.

Vlastos, Gregory (1984): 'Justice and Equality'. In: Waldron, Jeremy (ed.): *Theories of Rights*. Oxford: Oxford University Press.

Waldron, Jeremy (1992): 'Superseding Historic Injustice'. In: *Ethics*, Vol. 103 (1), pp. 4–28.

——— (1984): 'Introduction'. In: Waldron, Jeremy (ed.): *Theories of Rights*. Oxford: Oxford University Press, pp. 1–20.

Wallack, Michael (2006): 'Justice between Generations: The Limits of Procedural Justice'. In: Tremmel, Jörg C. (ed.): *Handbook of Intergenerational Justice*. Cheltenham: Edward Elgar Publishing, pp. 86–105.

Wanderwitz, Marco/Friedrich, Peter/Lührmann, Anna/Kauch, Michael (2008): 'Changing the German Constitution in Favor of Future Generations – Four Perspectives from the Young Generation'. In: Tremmel, Jörg (ed.) (2008): *Demographic Change and Intergenerational Justice. The Implementation of Long-term Thinking in Political Decision-Making*. Berlin/Heidelberg: Springer Verlag, 163–173.

Wenar, Leif (2011): 'Rights'. In: *Stanford Encyclopedia of Philosophy*. http://plato.stanford.edu/entries/rights/. Accessed 5 March 2012.

Weston, Burns H./Bach, Tracy (2009): 'Recalibrating the Law of Humans with the Laws of Nature: Climate Change, Human Rights, and Intergenerational Justice'. http://www.vermontlaw.edu/x4128.xml. Accessed 13 April 2011.

Westra, Laura (2006): *Environmental Justice and the Rights of Unborn and Future Generations*. London: Earthscan.

Wissenschaftlicher Beirat der Bundesregierung Globale Umweltveränderungen (WBGU) (2009): *Kassensturz für den Weltklimavertrag – der Budgetansatz*. Sondergutachten. Berlin.

Wolf, Clark (2008): 'Justice and Intergenerational Debt'. In: *Intergenerational Justice Review*, Vol. 8 (1), pp. 13–17.

Woodward, James (1986): 'The Non-Identity Problem'. In: *Ethics*. Vol. 96, 4/1986, pp. 804–831.

Worldbank (2012): 'Turn down the heat. Why a 4 degree warmer world should be avoided'. Washington. http://climatechange.worldbank.org/sites/default/files/Turn_Down_the_heat_Why_a_4_degree_centrigrade_warmer_world_must_be_avoided.pdf. Accessed 15 January 2013.

World Nuclear Association (2010): Nuclear Power in Germany. http://world-nuclear.org/info/inf43.html, Accessed 10 September 2011.

INDEX

www.ingramcontent.com/pod-product-compliance
Ingram Content Group UK Ltd.
Pitfield, Milton Keynes, MK11 3LW, UK
UKHW020206290926
15470UKWH00001B/29

9 781780 763637